# SWITCHING CIRCUIT AND LOGIC DESIGN

# SWITCHING CIRCUIT AND LOGIC DESIGN

Er. N.B.L. Mathur

ANMOL PUBLICATIONS PVT. LTD.
NEW DELHI-110 002 (INDIA)

**ANMOL PUBLICATIONS PVT. LTD.**
**Regd. Office:** 4360/4, Ansari Road, Daryaganj,
New Delhi-110002 (India)
Tel.: 23278000, 23261597, 23286875, 23255577
Fax: 91-11-23280289
*Email:* anmolpub@gmail.com
*Visit us at:* www.anmolpublications.com

**Branch Office:** No. 1015, Ist Main Road, BSK IIIrd Stage
IIIrd Phase, IIIrd Block, Bangalore-560 085 (India)
Tel.: 080-41723429 • Fax: 080-26723604
*Email:* anmolpublicationsbangalore@gmail.com

*Switching Circuit and Logic Design*



First Edition, 2011

ISBN 978-81-261-4486-0

PRINTED IN INDIA

---

Printed at Mehra Offset Press, Delhi.

# Contents

# Preface

The textbook has been designed for the undergraduate students of Electrical and Electronics, Electronics and Communication, Computer Science, Electronics and Instrumentation, Information Technology and Electronics and Control Engineering. This book provides an accessible and practical treatment to many combinational and sequential circuits. Each topic has been discussed in sufficient depth to expose the fundamental principles, concepts, techniques which are necessary to understand the subject thoroughly.

The book discusses threshold gates and programmable logic devices (PLDs). In addition, it elaborates on flip-flops and shift registers. Each chapter includes several fully worked-out examples so that the students get a thorough grounding in related design concepts.

*Author*

# Chapter 1

# Introduction

## DIGITAL SYSTEM

While the binary numeration system is an interesting mathematical abstraction, we haven't yet seen its practical application to electronics. What makes binary numeration so important to the application of digital electronics is the ease in which bits may be represented in physical terms.

Because a binary bit can only have one of two different values, either 0 or 1, any physical medium capable of switching between two saturated states may be used to represent a bit. Consequently, any physical system capable of representing binary bits is able to represent numerical quantities, and potentially has the ability to manipulate those numbers. This is the basic concept underlying digital computing.

Electronic circuits are physical systems that lend themselves well to the representation of binary numbers. Transistors, when operated at their bias limits, may be in one of two different states: either cutoff (no controlled current) or saturation (maximum controlled current).

If a transistor circuit is designed to maximize the probability of falling into either one of these states (and not operating in the linear, or active, mode), it can serve as a physical representation of a binary bit. A voltage signal measured at the output of such a circuit may also serve as a representation of a single bit, a low voltage representing a binary "0" and a (relatively) high voltage representing a binary "1." Consider the following transistor circuit:

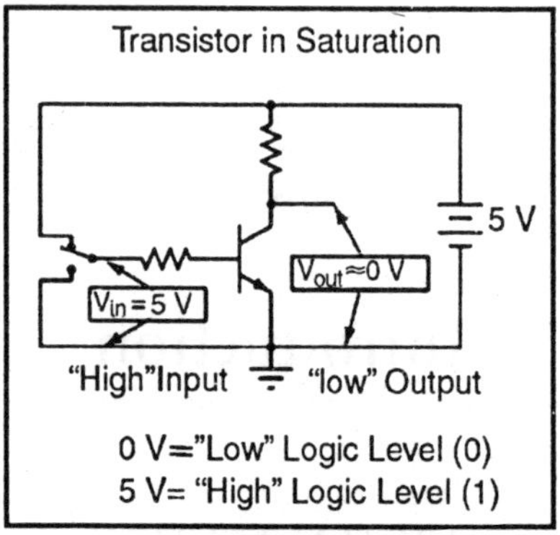


In this circuit, the transistor is in a state of saturation by virtue of the applied input voltage (5 volts) through the two-position switch. Because it's saturated, the transistor drops very little voltage between collector and emitter, resulting in an output voltage of (practically) 0 volts.

If we were using this circuit to represent binary bits, we would say that the input signal is a binary "1" and that the output signal is a binary "0." Any voltage close to full supply voltage (measured in reference to ground, of course) is considered a "1" and a lack of voltage is considered a "0." Alternative terms for these voltage levels are high (same as a binary "1") and low (same as a binary "0").

A general term for the representation of a binary bit by a circuit voltage is logic level. Moving the switch to the other position, we apply a binary "0" to the input and receive a binary "1" at the output:

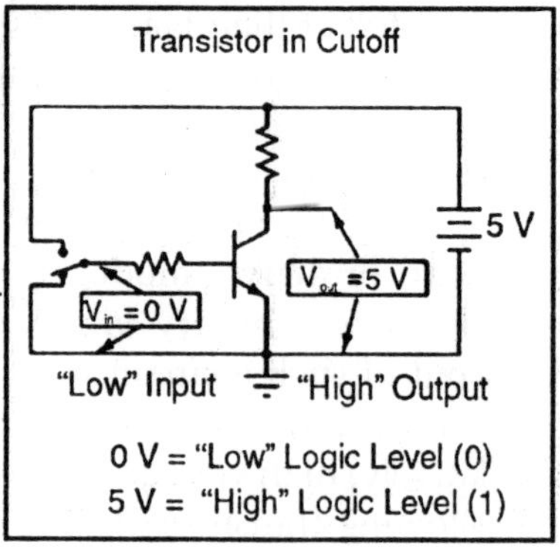

What we've created here with a single transistor is a circuit generally known as a logic gate, or simply gate. A gate is a special type of amplifier circuit designed to accept and generate voltage signals corresponding to binary 1's and 0's. As such, gates are not intended to be used for amplifying analog signals (voltage signals between 0 and full voltage).

Used together, multiple gates may be applied to the task of binary number storage (memory circuits) or manipulation (computing circuits), each gate's output representing one bit of a multi-bit binary number.

The gate shown here with the single transistor is known as an inverter, or NOT gate, because it outputs the exact opposite digital signal as what is input. For convenience, gate circuits are generally represented by their own symbols rather than by their constituent transistors and resistors. The following is the symbol for an inverter:

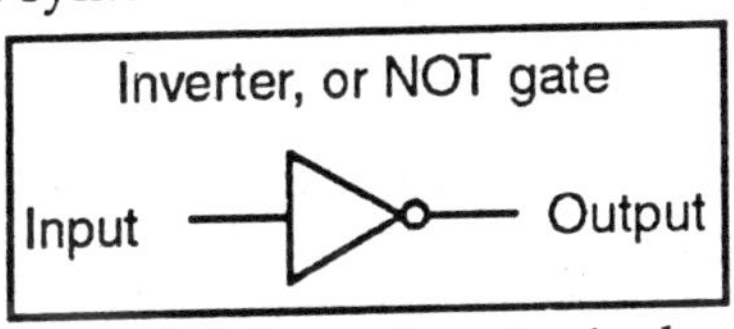


*An alternative symbol for an inverter is shown here:*

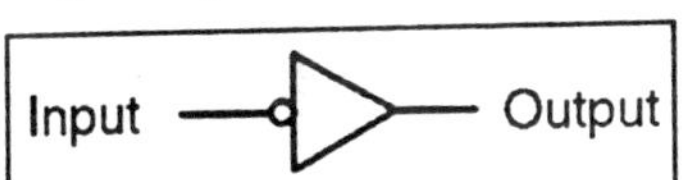


Notice the triangular shape of the gate symbol, much like that of an operational amplifier. As was stated before, gate circuits actually are amplifiers. The small circle, or "bubble" shown on either the input or output terminal is standard for representing the inversion function.

As you might suspect, if we were to remove the bubble from the gate symbol, leaving only a triangle, the resulting symbol would no longer indicate inversion, but merely direct amplification. Such a symbol and such a gate actually do exist, and it is called a buffer, the subject of the next section.

Like an operational amplifier symbol, input and output connections are shown as single wires, the implied reference point for each voltage signal being "ground." In digital gate circuits, ground is almost always the negative connection of a single voltage source (power supply).

Dual, or "split," power supplies are seldom used in gate circuitry. Because gate circuits are amplifiers, they require a source of power to operate. Like operational amplifiers, the power supply connections for digital gates are often omitted from the symbol for simplicity's sake. If we were to show all the necessary connections needed for operating this gate, the schematic would look something like this:

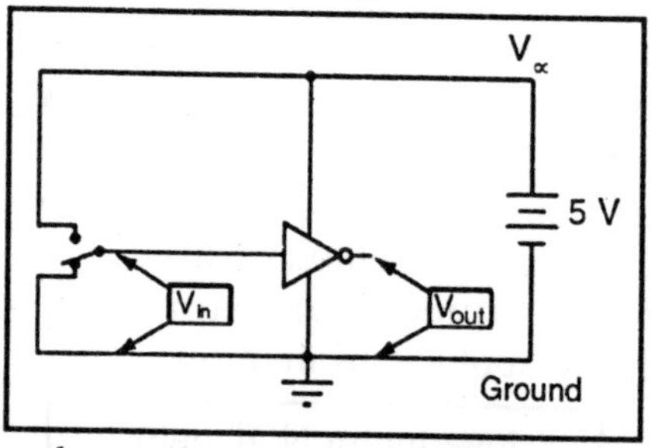


Power supply conductors are rarely shown in gate circuit schematics, even if the power supply connections at each gate are. Minimizing lines in our schematic, we get this:

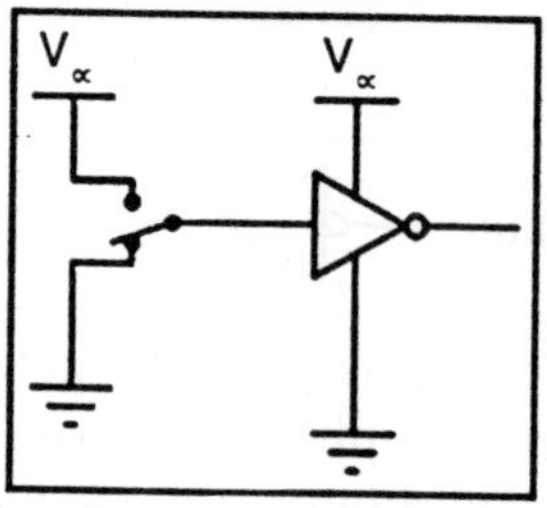

"$V_{cc}$" stands for the constant voltage supplied to the collector of a bipolar junction transistor circuit, in reference to ground. Those points in a gate circuit marked by the label "$V_{cc}$" are all connected to the same point, and that point is the positive terminal of a DC voltage source, usually 5 volts.

As we will see in other sections of this chapter, there are quite a few different types of logic gates, most of which have multiple input terminals for accepting more than one signal. The output of any gate is dependent on the state of its input(s) and its logical function.

One common way to express the particular function of a gate circuit is called a truth table. Truth tables show all combinations of input conditions in terms of logic level states (either "high" or "low," "1" or "0," for each input terminal of the gate), along with the corresponding output logic level, either "high" or "low." For the inverter, or NOT, circuit just illustrated, the truth table is very simple indeed:

NOT Gate Truth Table

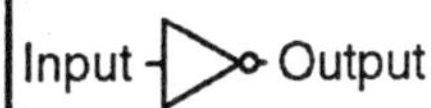


| Input | Output |
|---|---|
| 0 | 1 |
| 1 | 0 |

Truth tables for more complex gates are, of course, larger than the one shown for the NOT gate. A gate's truth table must have as many rows as there are possibilities for unique input combinations. For a single-input gate like the NOT gate, there are only two possibilities, 0 and 1. For a two input gate, there are four possibilities (00, 01, 10, and 11), and thus four rows to the corresponding truth table.

For a three-input gate, there are eight possibilities (000, 001, 010, 011, 100, 101, 110, and 111), and thus a truth table with eight rows are needed. The mathematically inclined will realise that the number of truth table rows needed for a gate is equal to 2 raised to the power of the number of input terminals.

## HIGH VACUUM DIODE

A vacuum diode is an electron valve that supports current flow through it in only one direction. It is composed of two electrical "parts" or elements and the evacuated glass envelope. Additionally, there is usually a heater, a coil of special alloy wire that can be driven to red heat by a modest application of voltage. (A vacuum diode without a heater would be a cold cathode diode.) The heater is coiled inside a cylindrical element called the cathode. The cathode is composed of a special alloy, or, more properly is coated with an alloy, that has high thermal emission characteristics. That means that if it's heated sufficiently, electrons actually leave the metallic crystal structure of the metal (boil off) and hang around outside in the space nearby forming what is called a space charge.

When we built the tube, we pumped all the air out, sealed the glass, and then fired a "getter" inside the tube to chemically bind the remaining bit of gas that was there. The tube really is highly evacuated.

That's important because we don't want all those light weight electrons bumping into gas atoms and molecules when they are hanging around in the space charge, and especially when transiting the gap between the cathode and the other element inside, the anode or plate.

If a voltage is applied across the cathode and plate, and the plate is negative, no current will flow in the tube. Electrons flow from negative to positive, but, darn it, they can't leave the plate to go to the cathode. That's called reverse biased. It's hooked up "backwards" like that. But if the cathode is connected to negative and the positive connected to the plate, electrons that were in the space charge will be driven away from the cathode and attracted to the plate. Current will flow in this direction. That's forward bias.

The hot cathode emits electrons in droves. Many more than are needed. They build up in the volume around the cathode and form a large pool of electrons just waiting for something positive to happen.

This pool of electrons is called the space charge. When the plate (anode) is made positive some of the electrons are attracted to it. They impact on it, are absorbed into the metal and electrons flow out of the plate connection into the battery or what ever provided the positive voltage. The other end of the voltage source must be connected back to the cathode in some way.

If the polarity is reversed which makes the plate negative with respect to the cathode the electrons in the space charge are repelled away from the plate and no conduction takes place. The plate is cold and is made of a metal that is a very poor emitter of electrons. The voltages are not nearly as high as those used by Crooks so there is no cold emission from the plate. This makes the diode conduct current in only one direction.

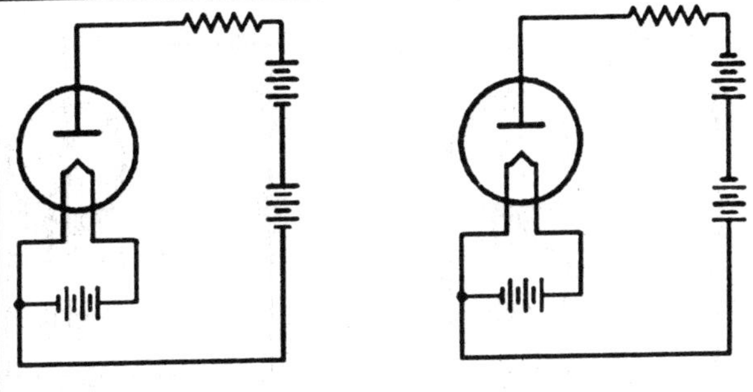

Heavy Conduction. No Conduction.

The cathode of a tube is coated with a very special material that emits electrons very easily. It can readily be damaged by operating above or below the proper temperature or by being "poisoned" by the residual air in the tube. The space charge serves another vital function.

There is no such thing as a perfect vacuum. In fact the space in near earth orbit is a better vacuum than can be made on the surface of earth. Farther out in space the vacuum is even better.

What this means is there are lots of air molecules running around inside of a vacuum tube. The space charge protects the cathode from these air molecules. Some of these molecules lose an electron and become positive ions. They are attracted to the negative cathode and if it weren't for the space charge they would impact on the surface of the cathode doing considerable damage.

Over time the ability of the cathode to emit electrons would be seriously impaired rendering the tube useless. As these ions run through the space charge they gain electrons from the many collisions and become negative ions. Their direction is reversed turning them away from the delicate cathode.

## INDIRECTLY HEATED CATHODES

Most vacuum tubes have a basically cylindrical configuration. Figure two shows how a filament (directly heated) cathode type diode is made along with a heater type (indirectly heated) cathode type diode.

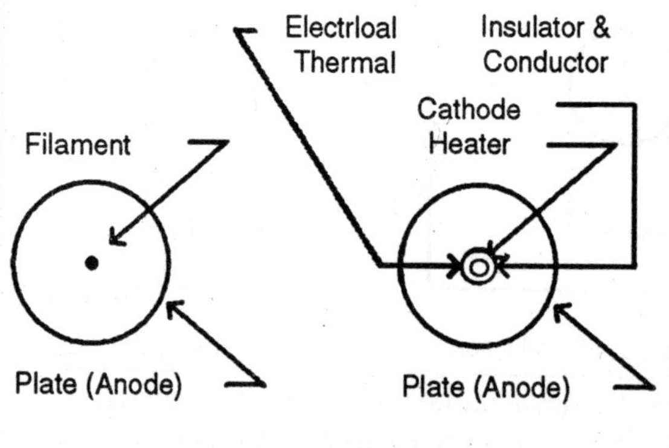


End View of Cylindrical Constructs.

## SEMICONDUCTOR DIODE

## INTRODUCTION

If we join a section of N-type semiconductor material with a similar section of P-type semiconductor material, we obtain a device known as a PN JUNCTION. (The area where the N and P regions meet is appropriately called the junction.) The usual characteristics of this device make it extremely useful in electronics as a diode rectifier.

The diode rectifier or PN junction diode performs the same function as its counterpart in electron tubes but in a different way.

The diode is nothing more than a two-element semiconductor device that makes use of the rectifying properties of a PN junction to convert alternating current into direct current by permitting current flow in only one direction. The schematic symbol of a PN junction diode is shown in figure below.

The vertical bar represents thecathode (N-type material) since it is the source of electrons and the arrow represents the anode. (P-type material) since it is the destination of the electrons.

The label "CR1" is an alphanumerical code used to identify the diode. In this figure, we have only one diode so it is labeled CR1 (crystal rectifier number one). If there were four diodes shown in the diagram, the last diode would be labeled CR4.

The heavy dark line shows electron flow. Notice it is against the arrow. For further clarification, a pictorial diagram of a PN junction and an actual semiconductor (one of many types) are also illustrated.

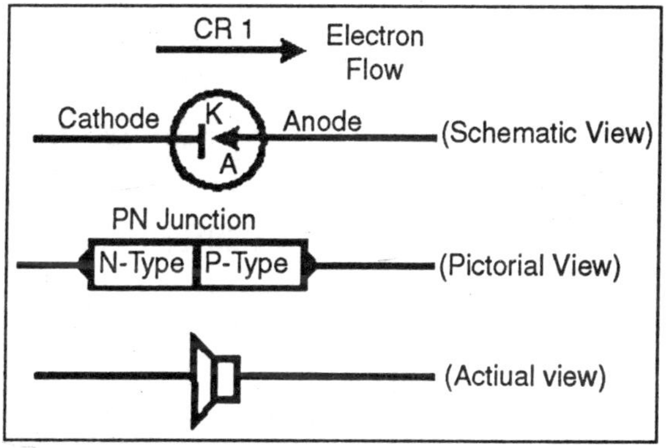


## CONSTRUCTION

Merely pressing together a section of P material and a section of N material, however, is not sufficient to produce a rectifying junction. The semiconductor should be in one piece to form a proper PN junction, but divided into a P-type impurity region and an N-type impurity region. This can be done in various ways.

One way is to mix P-type and N-type impurities into a single crystal during the manufacturing process. By so doing, a P-region is grown over part of a semiconductor's length and N-region is grown over the other part. This is called a GROWN junction and is illustrated in view A of figure below..

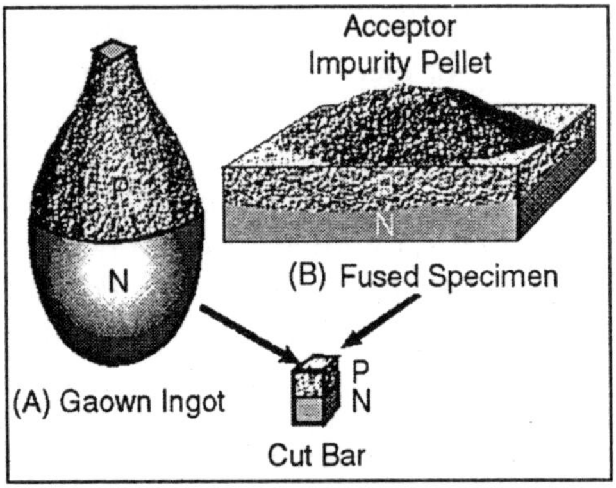


Another way to produce a PN junction is to melt one type of impurity into a semiconductor of the opposite type impurity. For example, a pellet of acceptor impurity is placed on a wafer of N-type germanium and heated. Under controlled temperature conditions, the acceptor impurity fuses into the wafer to form a P-region within it, as shown in view B of figure above. This type of junction is known as an ALLOY or FUSED-ALLOY junction, and is one of the most commonly used junctions. In figure shown below, a POINT-CONTACT type of construction is shown.

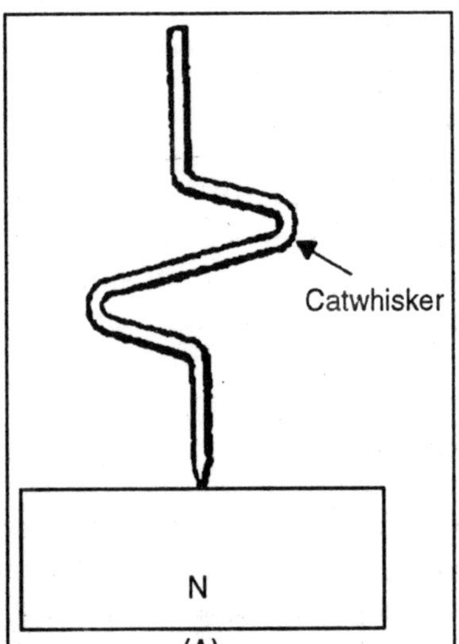

It consists of a fine metal wire, called a cat whisker, that makes contact with a small area on the surface of an N-type semiconductor as shown in view A of the figure. The PN union is formed in this process by momentarily applying a high-surge current to the wire and the N-type semiconductor. The heat generated by this current converts the material nearest the point of contact to a P-type material (view B).

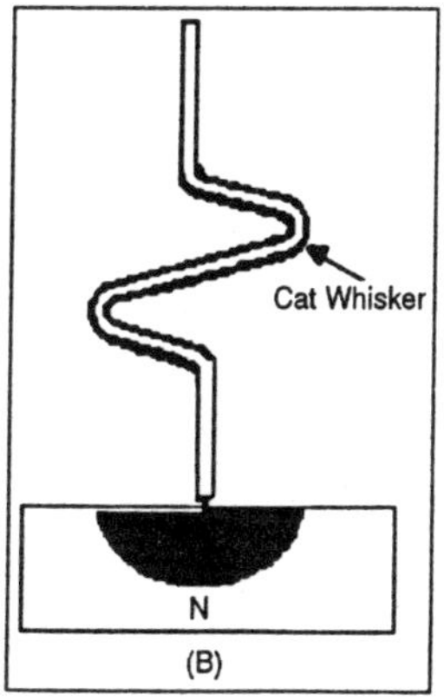


## TYPES OF DIODES

The first diode in figure below is a semiconductor diode which could be a small signal diode of the 1N914 type commonly used in switching applications, a rectifying diode of the 1N4004 (400V 1A) type or even one of the high power, high current stud mounting types. You will notice the straight bar end has the letter "k", this denotes the "cathode" while the "a" denotes anode. Current can only flow from anode to cathode and not in the reverse direction, hence the "arrow" appearance. This is one very important property of diodes.

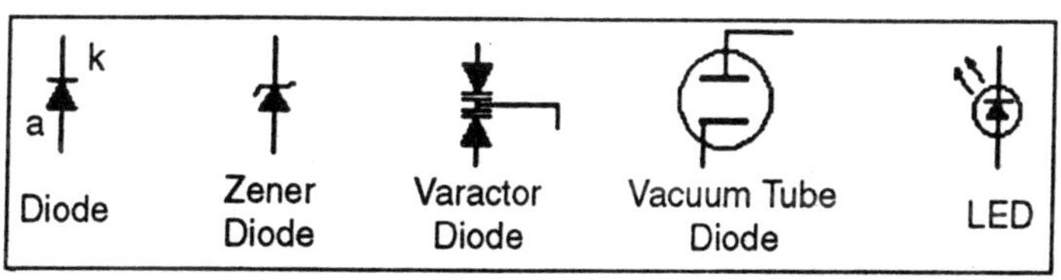


The second of the diodes is a zener diode which are fairly popular for the voltage regulation of low current power supplies. Whilst it is possible to obtain high current zener diodes, most regulation today is done electronically with the use of dedicated integrated circuits and pass transistors.

The next of the diodes in the schematic is a varactor or tuning diode. Depicted here is actually two varactor diodes mounted back to back with the DC control voltage applied at the common junction of the cathodes.

These cathodes have the double bar appearance of capacitors to indicate a varactor diode. When a DC control voltage is applied to the common junction of the cathodes, the capacitance exhibited by the diodes (all diodes and transistors exhibit some degree of capacitance) will vary in accordance with the applied voltage.

A typical example of a varactor diode would be the Philips BB204G tuning diodes of which there are two enscapsulated in a TO-92 transistor package. At a reverse voltage Vr (cathode to anode) of 20V each diode has a capacitance of about 16 pF and at Vr of 3V this capacitance has altered to about 36 pF. Being low cost diodes, tuning diodes have virtually replaced air variable capacitors in radio applications today.

The next diode is the simplest form of vacuum tube or valve. It simply has the old cathode and anode. These terms were passed on to modern solid state devices. Vacuum tube diodes are mainly only of interest to restorers and tube enthusiasts.

The last diode depicted is of course a light emitting diode or LED. A led actually doesn't emit as much light as it first appears, a single LED has a plastic lens installed over it and this concentrates the amount of light. Seven LED's can be arranged in a bar fashion called a seven segment LED display and when decoded properly can display the numbers 0 - 9 as well as the letters A to F.

## RECTIFYING DIODES

The principal early application of diodes was in rectifying 50 / 60 Hz AC mains to raw DC which was later smoothed by choke transformers and / or capacitors. This procedure is still carried out today and a number of rectifying schemes for diodes have evolved, half wave, full wave and bridge rectifiers.

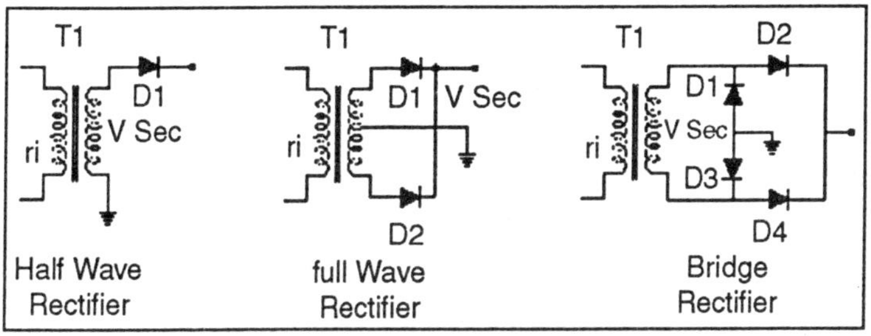


**Fig.** Rectifying Diodes

As examples in these applications the half wave rectifier passes only the positive half of successive cycles to the output filter through D1. During the negative part of the cycle D1 does not conduct and no current flows to the load. In the full wave application it essentially is two half wave rectifiers combined and because the transformer secondary is centre tapped, D1 conducts on the positive half of the cycle while D2 conducts on the negative part of the cycle. Both add together. This is more efficient. The full wave bridge rectifier operates essentially the same as the full wave rectifier but does not require a cetre tapped transformer. Further discussion may be seen on the topic power supplies.

A further application of rectifying diodes is in the conversion or detection of rf modulated signals to audio frequencies. Typical examples are am modulated signals being detected and early detection schemes for fm also used diodes for detecting modulation.

## VOLTAGE REGULATING DIODES

For relatively light current loads zener diodes are a cheap solution to voltage regulation. Zener diodes work on the principle of essentially a constant voltage drop at a predetermined voltage (determined during manufacture). An example is a Philips BZX79C12 type with a regulation range between 11.4V and 12.7V but typically 12V and a total power dissipation of 500 mW in a DO-35 package. The dissipation can be extended by using a series pass transistor, see power supplies. Notice in figure 3 there is a resistor to miminmise current drawn but mainly as an aid to dropping the supply voltage and reducing the burden on the zener diodes.

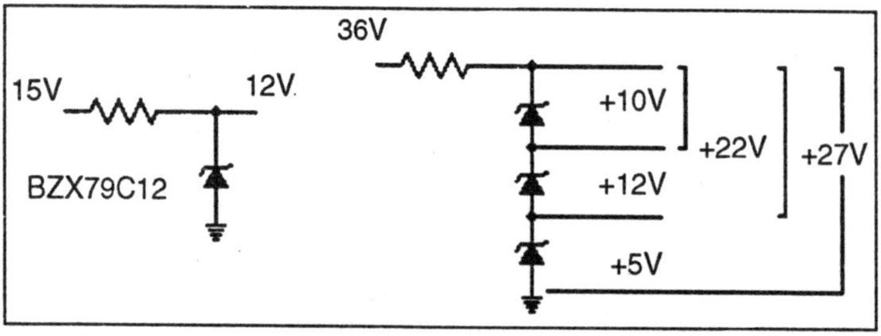


**Fig.** Zener Voltage Regulation Diodes

In the second schematic of figure above we have three zener diodes in series providing voltages of 5V, 10V, 12V, 22V and 27V all from a 36V supply. This configuration is not necessarily recommended especially when the current being drawn is seriously mismatched between voltages. It is presented purely out of interest.

## VARACTOR OR TUNING DIODES

These types of diodes work on the principle that all diodes exhibit some capacitance. Indeed the zener diode BZX79C12 quoted above has, according to the data book, a capacitance of 65-85 pF at 0V and measured at 1 Mhz.

For AM Radio band applications a specific diode has been devised. The Philips BB212 in a TO-92 case is one such type. Each of the diodes has a capacitance of 500 - 620 pF at a reverse bias of 0.5V and <22 pF at 8V. This diode's capacitance ratio is quoted at 22.5:1 which could not be achieved easily if at all with an air variable capacitor. This type of diode is depicted in figure 1 above. Several obvious advantages come immediately to mind, a small transistor type package, very low cost, ease of construction on a circuit board, can be mounted away from heat generating devices, frequency determining circuitry entirely dependent upon resistor values and ratios, DC voltage control can be either from frequency synthesiser circuits or perhaps a multi-turn potentiometer. Such a potentiometer aids band spreading and fine tuning if two potentiometers are used. The only real limitation is your imagination and the calculations involved.

## DIODES AS FREQUENCY MULTIPLIERS

Just one more example of the versatility of diodes is the frquency doubling circuit depicted in Figure. Now if that looks a lot like the full wave rectifier from figure above you would be correct. That is why the ripple frequency for 50 / 60 Hz always comes out at 100 / 120 Hz.

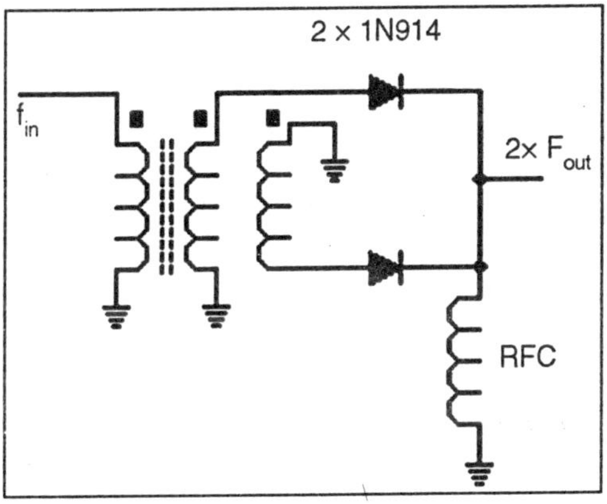


**Fig.** Diodes as Frequency Multipliers

Here the input is a wide band transformer and the signal passes to a full wave rectifier comprising two 1N914 diodes. The DC component caused by the rectification passes to ground through RFC which of course presents a high impedance to the rf porion of the signal but essentially a short circuit for DC. The original signal should be down about 40 dB and with this type of circuit there would be a loss of somewhere around 7.5 dB so the 2 × signal would require further amplification to restore that loss.

## DIODES AS MIXERS

With some subtle re-arrangement to figure 4 we can get

the circuit to function as a two diode frequency mixer. Note that there are other diode arrangements as well in this application.

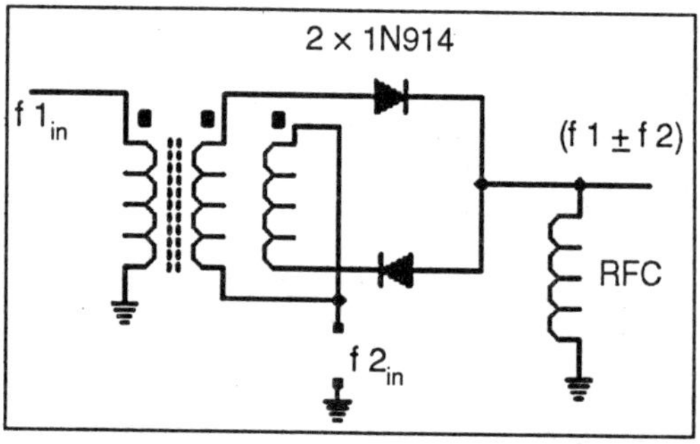


**Fig.** Diodes as Frequency Mixers

The diodes here act as switches and it can be mathematically shown that only the sum and difference signals will result. For example, if F1 was 5 Mhz and F2 was 3 Mhz then the sum and difference signals from the diodes would be 8 Mhz and 2 Mhz. None of the original signals appear at the output and this is a most important property of using diodes as mixers.

It should be noted that although 1N914 diodes are depicted you would normally use hot carrier diodes in any serious application and the diodes need to be well matched.

## Applications of Switching Diodes

Similar types of diodes have been developed specifically for band switching purposes. Although a typical 1N914 type switching diode can be used for such purposes it is preferable to use diodes which have been optimised for such purposes because the Rd on is much lower. This means the diode resistance Rd can have a serious affect on rf circuits in particular the "Q" of a tuned circuit. One example of a low Rd device is the Philips BA482 diode used for band switching in television tuners. It has a typical Rd of 0.4 ohms at a forward current of 10 mA.

In figure above we have one application where switching diodes operate. All diodes serve to switch in or out capacitors in the diagram which is presented here just to illustrate one single application of switching diodes, many, many more applications exist. Again the limit is your imagination.

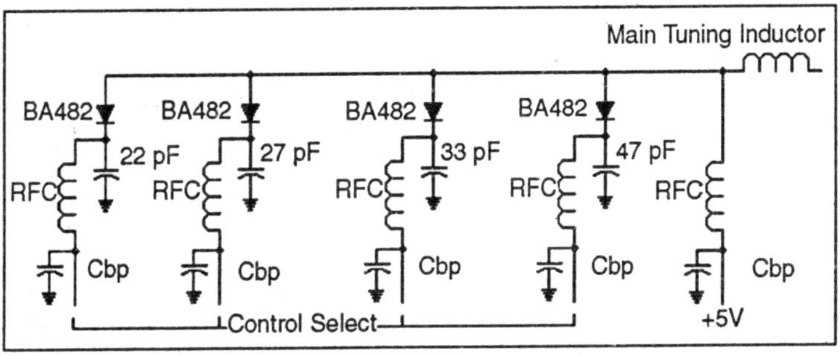


**Fig.** Applications of Switching Diodes

The switching diodes in figure above switchs in or out successively higher values of capacitors as each control select line is "grounded". The voltage from the +5V feed line proceeds through the diode at DC thus opening the diode and making it appear "transparent" for rf purposes. The capacitor with the value attached is then "switched" into circuit. Other components marked RFC and Cbp are chokes and bypass capacitors for "clean" switching. The bypass capacitors and choke values would be determined by the frequency of operation.

We could just have easily have switched inductors instead of capacitors. Note why Rd is quite important on overall circuit performance. If we were using inductors the diode resistance Rd would have a significant affect on inductor "Q" which in turn would affect filter performance, if it was in fact an LC filter application.

## SWITCHING DIODES IN LOGIC CIRCUITS

If you you completed the tutorial on digital basics you should be aware of binary numbers. There are a whole range of digital building blocks available and just by way of one illustration of using diodes we have presented the 74HC4040 twelve stage binary ripple counter (there are others with varying number of stages).

In the schematic of figure 7 we have this counter which divides by successive division of two for twelve stages. Initially because there is no voltage drop across the resistor a high appears on all anodes as well as on pin 4 the master reset causing the counter to reset forcing all outputs low and in turn a voltage drop across each diode and across the resistor and a low on reset. Progressively each of the outputs change from low to high for a certain period of time and without unduly complicating matters when all outputs as selected by our diode combination (in this particular case 1 + 2 + 32 + 64 = 99) are simultaneously high the voltage drop across the resistor will cease and cause pin 11 (reset which was formerly low) to go high and reset all the internal ripple counters.

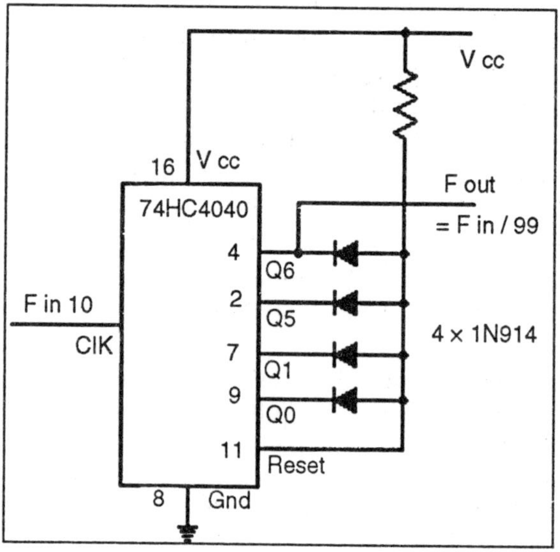


**Fig.** Applications of Switching Diodes in Digital Logic Circuits

At the same time pin 4 changes state also with reset. It can been shown this happens once every 99 periods. Simply by placing diodes on the right outputs we can select to divide by any number up to 4095 using this particular counter.

## LIGHT-EMITTING-DIODES OR LED'S

Many circuits use a led as a visual indicator of some sort

even if only as an indicator of power supply being turned on. A sample calculation of the dropping resistor is included in figure

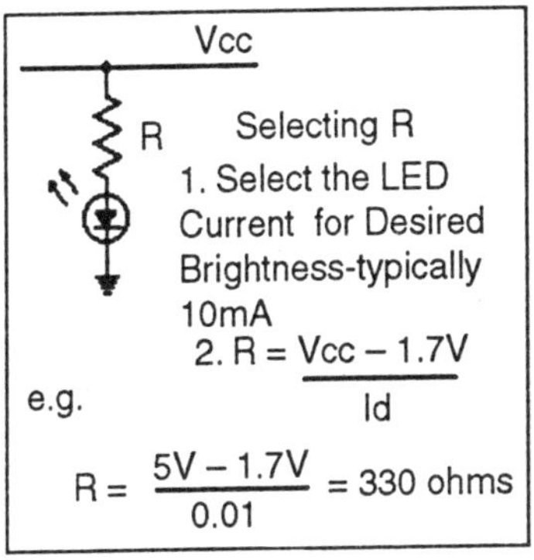


**Fig.** Connecting Light Emitting Diodes (LED's) to Supply

Most leds operate at 1.7V although this is not always the case and it is wise to check. The dropping resistor is simply the net of supply voltage minus the 1.7V led voltage then divided by the led brightness current expressed as "amps" (ohms law). Note the orientation of both cathode and anode with respect to the ground end and the supply end. Usually with a led the longer lead is the anode.

## IDEAL DIODE

## DIODES

Diodes are different and useful electrical components. Diodes are used in many applications like the following.

- Converting AC power from the 60Hz line into DC power for radios, televisions, telephone answering machines, computers, and many other electronic devices.
- Converting radio frequency signals into audible signals in radios.

Here are the goals for this lesson unit.

- Given a circuit with a diode,
  - Be able to use a simplified model of a diode to

predict when current flows through the diode, and when it does not.

– Be able to use information about current flowing to predict other behaviour in a circuit.

## IDEAL DIODE MODEL

The perfect diode model must have the following characteristics:

- It is a two-port device (at least external to the model/ subcircuit for the device).
- When the externally applied voltage from Cathode to Anode terminals is negative, essentially no internal current must flow (subject of course to numerical error limit considerations).
- When condition 2. is not met, the device current (and voltage) must be determined by the external circuitry, the device behaving as a short circuit (subject of course to the same numerical error limit considerations).

Using an internal opposing voltage source, one can construct a model and a test circuit for the first pass at an ideal diode as shown in Figure:

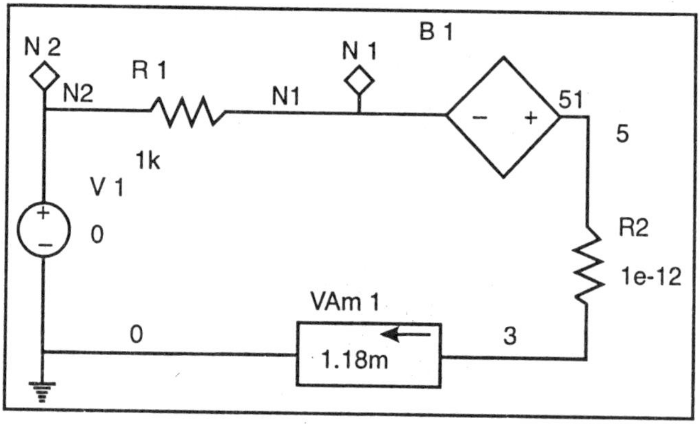


**Fig.** Ideal Diode Test Circuit

Performing a transient analysis on the circuit produces the results in Figure. The results are as expected, with the simulated ideal diode consisting of the 'B1' and 'R2' elements, having its anode at node 'N1" and its cathode at node '3'.

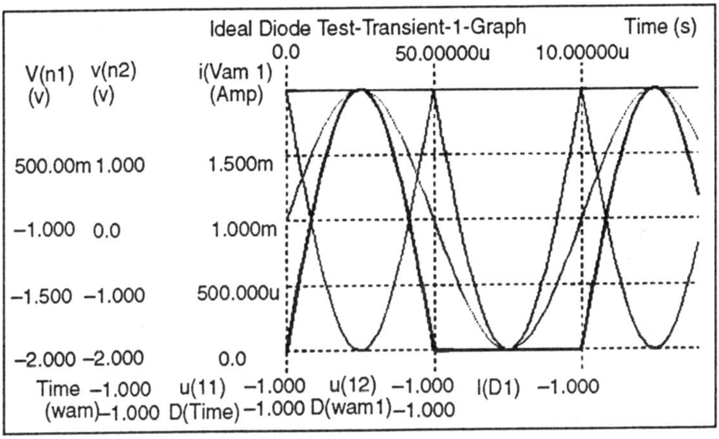


**Fig.** Ideal Diode Test Circuit

The operation of the diode circuit is straightforward. The applied voltage is opposed by the 'B1' voltage generator. The voltage of the 'B1' generator is the resultant of two levels. The first part is the absolute value of the voltage from node 'N1' to ground. The second is the uramp value of the voltage from node 'N1' to ground. The uramp value is the value of the 'N1' voltage when it is greater than zero.

Thus, during the half-sinusoid interval when the applied voltage is (or would be) positive, the voltage is unopposed. The equivalent diode 'conducts' with an 'ON' resistance equal to that of R2. During the second, negative half of the rectified sinusoid, the voltage generator 'B1' causes the voltage at 'N1' to be opposed by an essentially equal level, resulting in no appreciable current flow. This is of course an open circuit.

Note that the currents indicated by the ammeter and by the 'B1' source value are 180 degrees out of phase. This is caused by the sense of the 'B1' generator in the circuit. For a voltage source, within that source, 'positive' conventional current flows from negative to positive levels.

So far so good, however the sense of the 'B1' generator is negative when it is (for all essential purposes) non-zero. This causes the 'B1' current to seem to be wrong. However as this will be external to the diode model/subcircuit it is of no consequence.

This ideal diode could be added to the library as a device, perhaps as a symbol of a normal diode with a letter 'I' beside it to denote it as an ideal diode.

This is left as an exercise for the reader. This device can be associated with series resistance, parallel resistance and or capacitance as needed to make it somewhat less ideal and ease simulation difficulties at the transition points from conduction to non-conduction and the converse.

## SWITCHING DIODE AS HALF WAVE RECTIFIER

Now we come to the most popular application of the diode: *rectification*. Simply defined, rectification is the conversion of alternating current (AC) to direct current (DC). This involves a device that only allows one-way flow of electrons.

As we have seen, this is exactly what a semiconductor diode does. The simplest kind of rectifier circuit is the *half-wave* rectifier. It only allows one half of an AC waveform to pass through to the load.

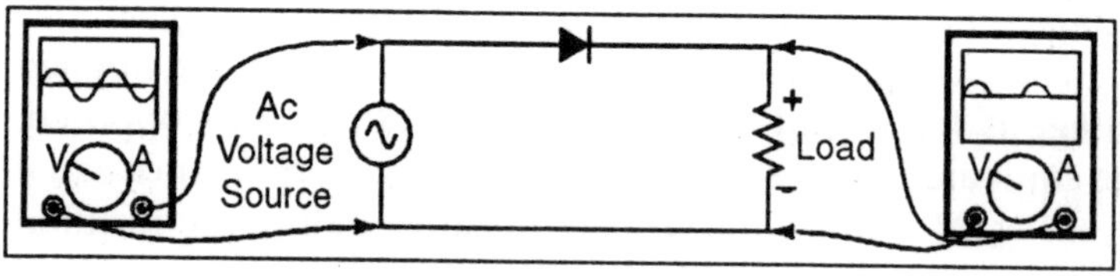


**Fig.** Half-wave Rectifier Circuit.

For most power applications, half-wave rectification is insufficient for the task. The harmonic content of the rectifier's output waveform is very large and consequently difficult to filter. Furthermore, the AC power source only supplies power to the load one half every full cycle, meaning that half of its capacity is unused.

Half-wave rectification is, however, a very simple way to reduce power to a resistive load. Some two-position lamp dimmer switches apply full AC power to the lamp filament for "full" brightness and then half-wave rectify it for a lesser light output.

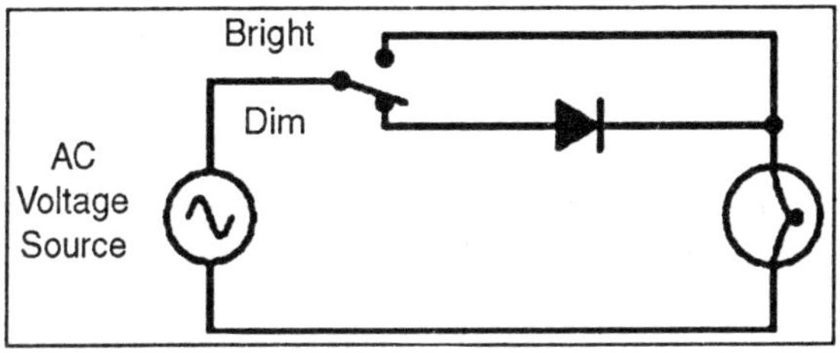


**Fig.** Half-wave Rectifier Application: Two Level Lamp Dimmer.

In the "Dim" switch position, the incandescent lamp receives approximately one-half the power it would normally receive operating on full-wave AC. Because the half-wave rectified power pulses far more rapidly than the filament has time to heat up and cool down, the lamp does not blink. Instead, its filament merely operates at a lesser temperature than normal, providing less light output.

This principle of "pulsing" power rapidly to a slow-responding load device to control the electrical power sent to it is common in the world of industrial electronics. Since the controlling device (the diode, in this case) is either fully conducting or fully nonconducting at any given time, it dissipates little heat energy while controlling load power, making this method of power control very energy-efficient. This circuit is perhaps the crudest possible method of pulsing power to a load, but it suffices as a proof-of-concept application.

## SWITCHING DIODE AS VOLTAGE REGULATOR

A Zener Diode is an electronic component which can be used to make a very simple voltage regulator circuit. This circuit enables a fixed stable voltage to be taken from an unstable voltage source such as thebattery bank of a renewable energy system which will fluctuate depending on the state of charge of the bank.

## ZENER DIODE VOLTAGE REGULATOR CIRCUIT

Pictured above is a very simple voltage regulator circuit

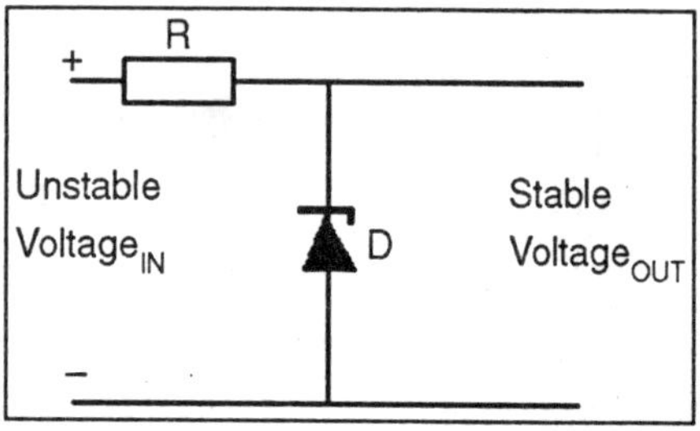


requiring just one zener diode (available from the REUK Shop) and one resistor. As long as the input voltage is a few volts more than the desired output voltage, the voltage across the zener diode will be stable.

As the input voltage increases the current through the Zener diode increases but the voltage drop remains constant - a feature of zener diodes.

Therefore since the current in the circuit has increased the voltage drop across the resistor increases by an amount equal to the difference between the input voltage and the zener voltage of the diode.

## SWITCHING DIODE AS FULL WAVE RECTIFIER

In a Full-wave rectifier circuit two diodes are now used, together with a transformer whose secondary winding is split equally into two and has a common centre tapped connection, (C). Now each diode conducts in turn when its Anode terminal is positive with respect to the centre point C as shown below.

### FULL-WAVE RECTIFIER CIRCUIT

The circuit consists of two *Half-wave* rectifiers connected to a single load resistance with each diode taking it in turn to supply current to the load. When point A is positive with respect to point B, diode $D_1$conducts in the forward direction as indicated by the arrows.

When point B is positive (in the negative half of the cycle) with respect to point A, diode $D_2$ conducts in the forward direction and the current flowing through resistor R is in the same direction for both circuits.

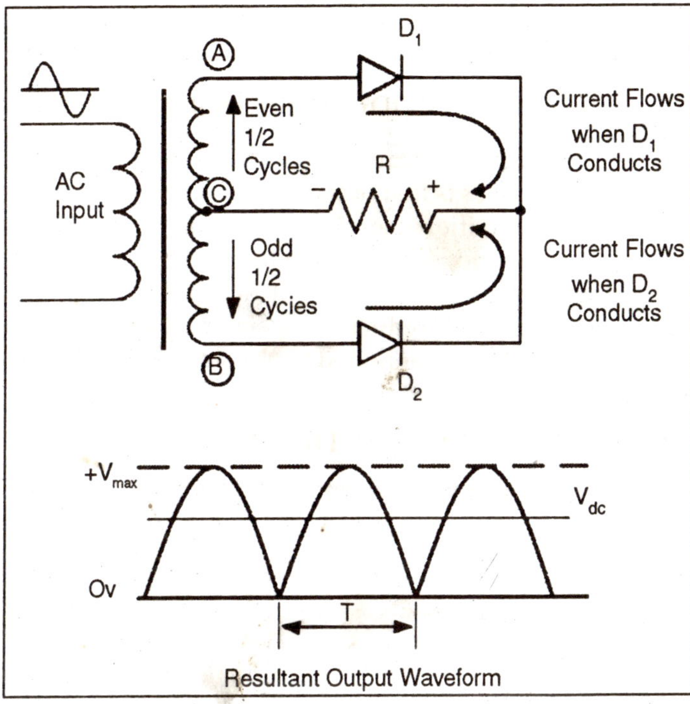


As the output voltage across the resistor R is the sum of the two waveforms, this type of circuit is also known as a "bi-phase" circuit.

As the spaces between each half-wave developed by each diode is now being filled in by the other diode the average DC output voltage across the load resistor is now double that of the single half-wave rectifier circuit and is about 0.637$V_{max}$ of the peak voltage, assuming no losses.

$$V_{d.c.} = \frac{V_{max}}{\pi} = 0.637V_{max} = 0.9V_S$$

The peak voltage of the output waveform is the same as before for the half-wave rectifier provided each half of the transformer windings have the same rms voltage value. To obtain a different d.c. voltage output different transformer ratios can be used, but one main disadvantage of this type of rectifier is that having a larger transformer for a given power output with two separate windings makes this type of circuit costly compared to a "Bridge Rectifier" circuit equivalent.

## SWITCHING DIODE AS BRIDGE RECTIFIER

Another type of circuit that produces the same output as a full-wave rectifier is that of the Bridge Rectifier. This type of single phase rectifier uses 4 individual rectifying diodes connected in a "bridged" configuration to produce the desired output but does not require a special centre tapped transformer, thereby reducing its size and cost. The single secondary winding is connected to one side of the diode bridge network and the load to the other side as shown below.

### THE DIODE BRIDGE RECTIFIER

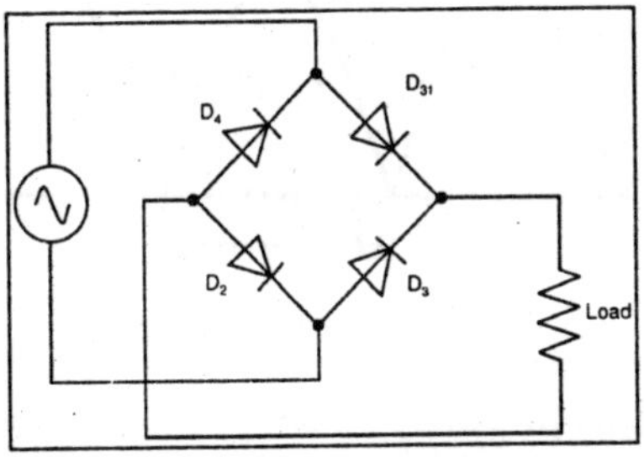


The 4 diodes labelled $D_1$ to $D_4$ are arranged in "series pairs" with only two diodes conducting current during each half cycle. During the positive half cycle of the supply, diodes D1 and D2 conduct in series while diodes D3 and D4 are reverse biased and the current flows through the load as shown below.

### The Positive Half-cycle

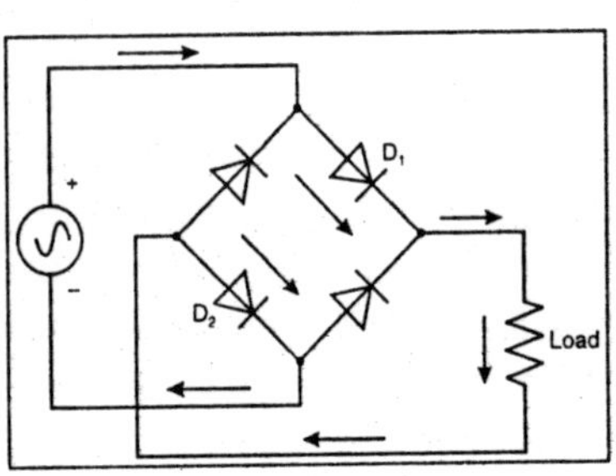

During the negative half cycle of the supply, diodes D3 and D4 conduct in series, but diodes D1 and D2 switch of as they are now reverse biased. The current flowing through the load is the same direction as before.

### The Negative Half-cycle

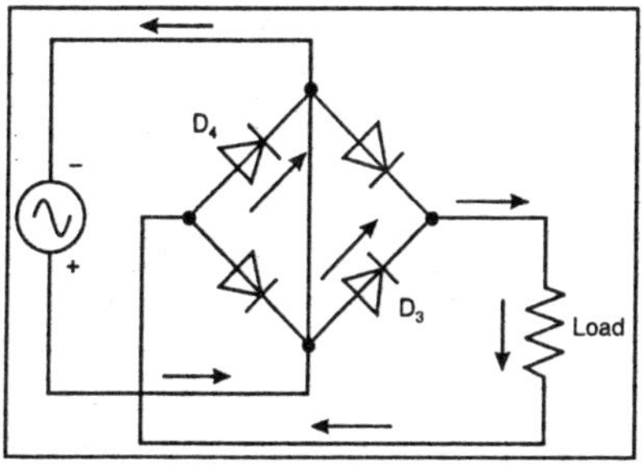


As the current flowing through the load is unidirectional, so the voltage developed across the load is also unidirectional the same as for the previous two diode full-wave rectifier, therefore the average DC voltage across the load is $0.637V_{max}$ and the ripple frequency is now twice the supply frequency (e.g. 100Hz for a 50Hz supply).

## SMOOTHING CAPACITOR

We saw in the previous section that the single phase half-wave rectifier produces an output wave every half cycle and that it was not practical to use this type of circuit to produce a steady DC supply. The full-wave bridge rectifier however, gives us a greater mean DC value (0.637 Vmax) with less superimposed ripple while the output waveform is twice that of the frequency of the input supply frequency. We can therefore increase its average DC output level even higher by connecting a suitable smoothing capacitor across the output of the bridge circuit as shown below.

### Full-wave Rectifier with Smoothing Capacitor

The smoothing capacitor converts the full-wave rippled output of the rectifier into a smooth DC output voltage. Two important parameters to consider when choosing a suitable a capacitor are its *Working Voltage,* which must be higher than the no-load output value of the rectifier and its *Capacitance Value,* which determines the amount of ripple that will appear superimposed on top of the DC voltage. Too low a value and the capacitor has little effect. As a general rule of thumb, we are looking to have a ripple voltage of less than 100mV peak to peak.

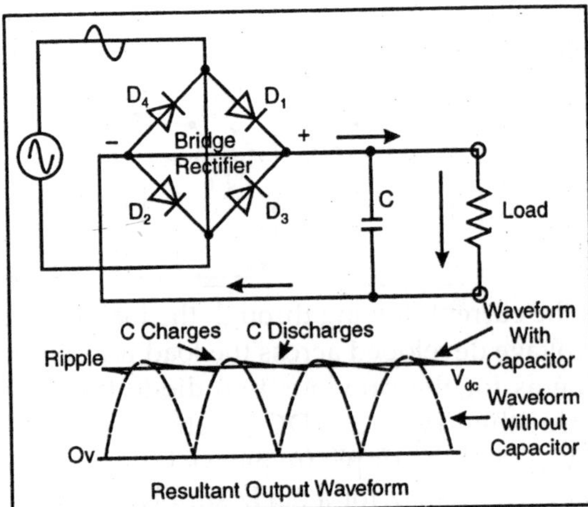


The main advantages of a full-wave bridge rectifier is that it has a smaller AC ripple value for a given load and a smaller reservoir or smoothing capacitor than an equivalent half-wave rectifier. Therefore, the fundamental frequency of the ripple voltage is twice that of the AC supply frequency (100Hz) where for the half-wave rectifier it is exactly equal to the supply frequency (50Hz).

The amount of ripple voltage that is superimposed on top of the DC supply voltage by the diodes can be virtually eliminated by adding a much improved $\pi$-filter (pi-filter) to the output terminals of the bridge rectifier.

This type of low-pass filter consists of two smoothing capacitors, usually of the same value and a choke or inductance across them to introduce a high impedance path to the alternating ripple component. Another more practical and cheaper alternative is to use a 3-terminal voltage regulator IC, such as a LM7805 which can reduce the ripple by more than

70dB (Datasheet) while delivering over 1amp of output current.

## INDUCTOR FILTER

INDUCTOR FILTERS use an inductor called a choke to filter the pulsating dc input. Because ofthe impedance offered to circuit current, the output of the filter is at a lower amplitude than the input

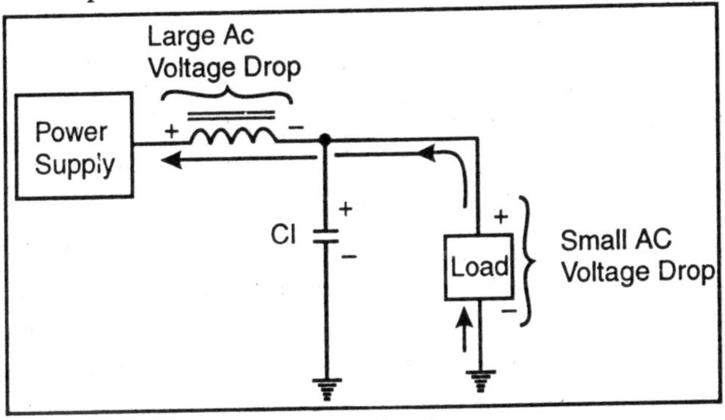


## CAPACITOR FILTER

The simple capacitor filter is the most basic type of power supply filter. The application of the simple capacitor filter is very limited. It is sometimes used on extremely high-voltage, low-current power supplies for cathode-ray and similar electron tubes, which require very little load current from the supply. The capacitor filter is also used where the power-supply ripple frequency is not critical; this frequency can be relatively high. The capacitor (C1) shown in figure 4-15 is a simple filter connected across the output of the rectifier in parallel with the load.

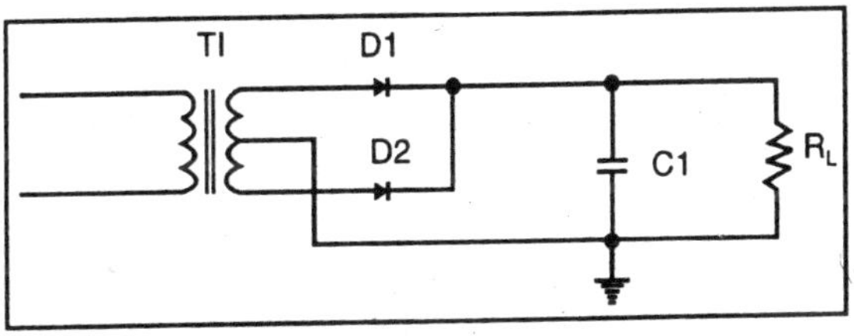


**Fig.** Full-wave Rectifier with a Capacitor Filter.

When this filter is used, the RC charge time of the filter capacitor (C1) must be short and the RC discharge time must be long to eliminate ripple action.

In other words, the capacitor must charge up fast, preferably with no discharge at all. Better filtering also results when the input frequency is high; therefore, the full-wave rectifier output is easier to filter than that of the half-wave rectifier because of its higher frequency.

For you to have a better understanding of the effect that filtering has on $E_{avg}$, a comparison of a rectifier circuit with a filter and one without a filter is illustrated in views A and B of figure above.

The output waveforms in figure 4-16 represent the unfiltered and filtered outputs of the half-wave rectifier circuit. Current pulses flow through the load resistance ($R_L$) each time a diode conducts.

The dashed line indicates the average value of output voltage. For the half-wave rectifier, $E_{avg}$ is less than half (or approximately 0.318) of the peak output voltage.

This value is still much less than that of the applied voltage. With no capacitor connected across the output of the rectifier circuit, the waveform in view A has a large pulsating component (ripple) compared with the average or dc component.

When a capacitor is connected across the output (view B), the average value of output voltage ($E_{avg}$) is increased due to the filtering action of capacitor C1.

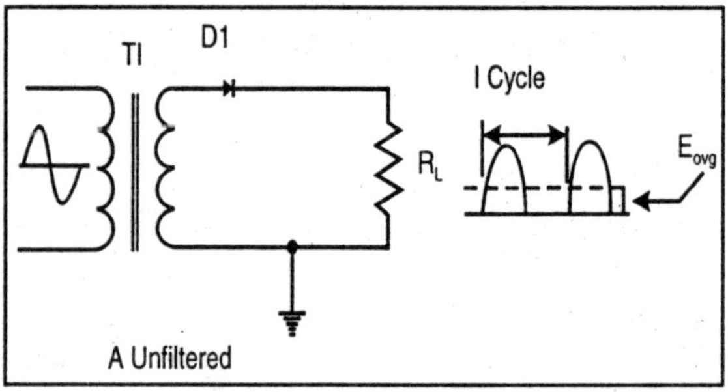


**Fig.** Half-wave Rectifier with and without Filtering.

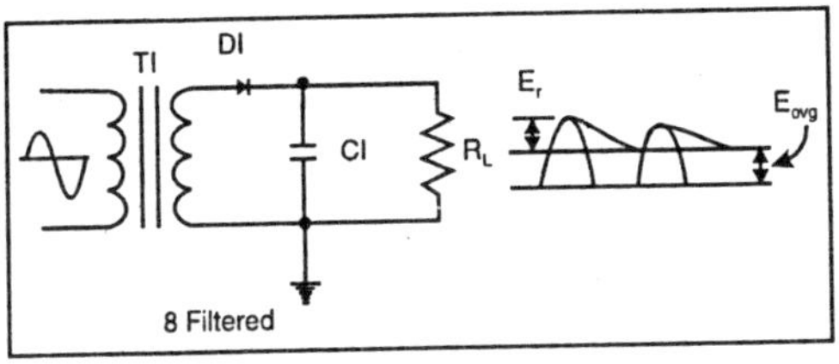


**Fig.** Half-wave Rectifier with and without Filtering.

The value of the capacitor is fairly large (several microfarads), thus it presents a relatively low reactance to the pulsating current and it stores a substantial charge.

# MOSFET

## INTRODUCTION

Mosfets are 4-terminal devices with a gate that controls the conduction of the drain to source channel. The body (or substrate) of the typical n-channel enhancement mosfet is a P-type material with the drain and source formed from N-type materials. The oxide layer insulates the metal gate connection from the other layers. There are holes in the oxide layer left for the metallic contacts of the drain and source.

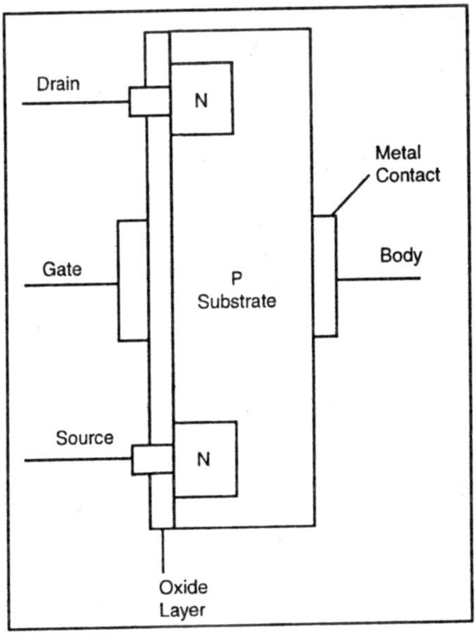

The metal oxide semiconductor field effect transistor (MOSFET) is one of the cornerstones of modern semiconductor technology.

The general structure is a lightly doped *p*-type substrate, into which two regions, the source and the drain, both of heavily doped *n*-type semiconductors have been embedded. The symbol *n*+ is used to denote this heavy doping.

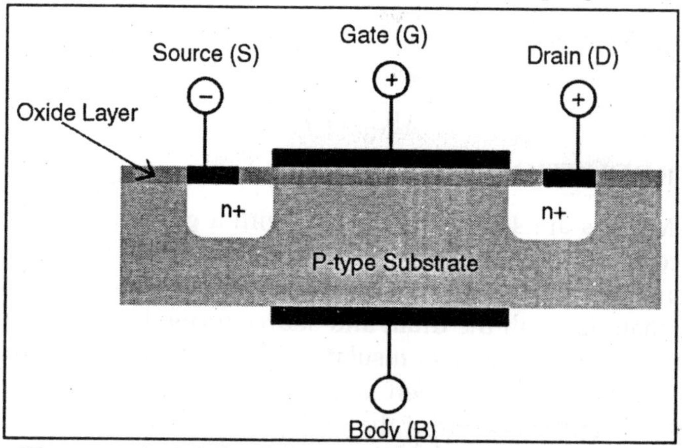


The source and the drain are about 1 ìm apart. Metallized contacts are made to both source and drain, generally using aluminium. The rest of the substrate surface is covered with a thin oxide film, typically about 0.05 ìm thick. The gate electrode is laid on top of the insulating oxide layer, and the body electrode in the above diagram provides a counter electrode to the Gate. The thin oxide film contains silicon dioxide ($SiO_2$), but it may well also contain silicon nitride ($Si_3N_4$) and silicon oxynitride ($Si_2N_2O$).

The *p*-type doped substrate is only very lightly doped, and so it has a very high electrical resistance, and current cannot pass between the source and drain if there is zero voltage on the gate. Application of a positive potential to the gate electrode creates a strong electric field across the *p*-type material even for relatively small voltages, as the device thickness is very small and the field strength is given by the potential difference divided by the separation of the gate and body electrodes.

## CONSTRUCTION

A MOSFET consists of an n-(p-)doped silicon substrate with two, highly p-(n-)doped contacts, source and drain. The so-called channel region inbetween is covered by an insulating layer, the gate-oxide, which is in contact with the gate electrode. Without applying a voltage at the gate electrode, no current can flow from source to drain as the pn-juctions between each contact and the substrate act as two opposite diodes. When applying a positive (negative) voltage at the gate electrode, the channel region close to the gate oxide is "inverted" (i.e. from n-(p-) to p-(n-)doped) and current can flow between source and drain.

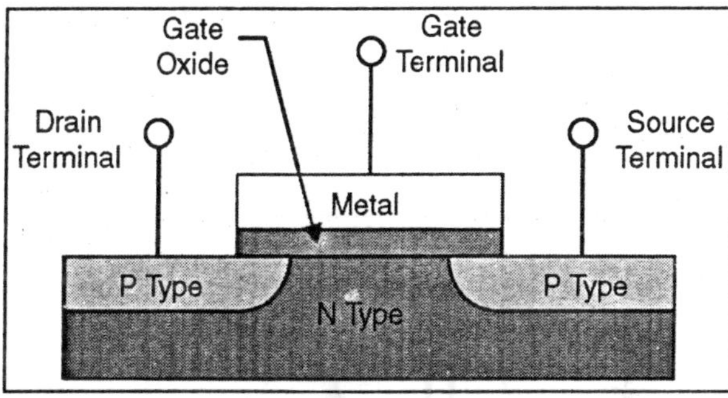


Central to the functionality is the thin insulating layer, the gate-oxide. The gate-oxide acts as the dielectricum of a capacitor which attracts charge carriers into the channel region. Up to now, silicon dioxide ($SiO_2$) has been used as a gate-oxide. The so far unparallelized electric and structural properties of silicon's native oxide - commonly known as window glass - are said to be one of the main reasons that silicon is today's semiconductor of choice.

Due to electrical considerations, the gate oxide thickness has to be proportional to the gate length (i.e. the distance between source and drain). The lateral scaling of transistor dimensions thus directly translates into a reduction of the gate-oxide thickness. Right now, oxide with a thickness of about 1.5 nm (one fifty-thousandth of the width of human hair) are employed. In 2007 less than 1 nm will be needed which corresponds to only 4 atomic layers. At these thick nesses, quantum mechanical tunneling currents through the oxide become intolerable.

## WORKING

N-P junctions act as diodes, and because the layers of the mosfet have P-type and N-type materials in close contact, diodes are formed from the body to the drain or source, as shown in the drawing here of the 4-terminal mosfet. If you were to leave the body connection floating, there could be no voltage flow through the two PN body diodes because they are reverse connected to each other.

However, in the 3-terminal mosfets (like the BS170) that are commonly used in guitar effects pedals, the body and source are connected internally and this shorts out the PN junction of the lower diode and removes it from the circuit. The top body diode can now conduct from drain to source because of the internal connection and therefore has to be taken into consideration during circuit design.

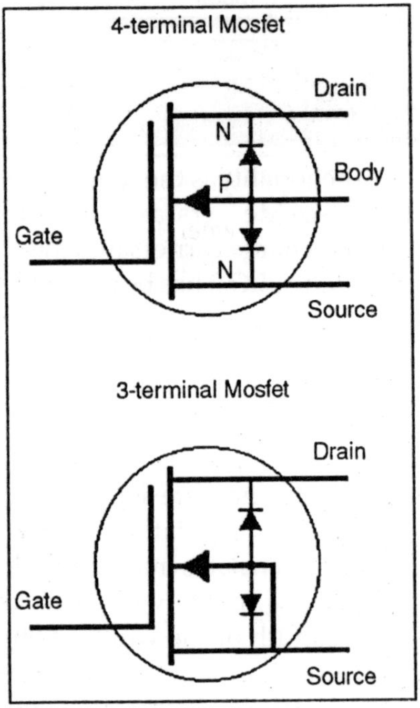

This means that every 3-terminal mosfet has a body diode due the construction of the device. It is not added as a separate component inside the mosfet.

However, there is something even more interesting in the 4-terminal mosfet. The drain-to-body-to-source structure is an N-P-N layered structure NPN, as in transistor. There is a bipolar transistor in every mosfet!

## BIPOLAR JUNCTION TRANSISTORS

A Bipolar Junction Transistor (a.k.a. a BJT or Bipolar Transistor) is an active semiconductor device formed by two P-N junctions whose function is amplification of an electric current. Bipolar transistors are made from 3 sections of semiconductor material (alternating P-type and N-type), with 2 resulting P-N junctions. Schematically, a bipolar transistor can be thought of in this fashion.

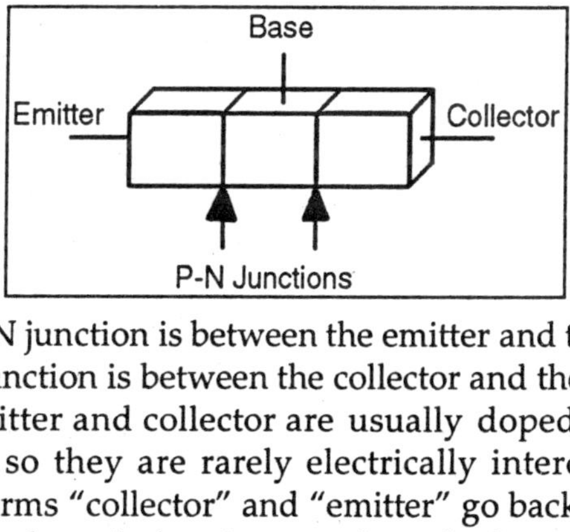


One P-N junction is between the emitter and the base; the other P-N junction is between the collector and the base. Note that the emitter and collector are usually doped somewhat differently, so they are rarely electrically interchangeable. While the terms "collector" and "emitter" go back to vacuum tube days, the base derives its name from the first point-contact transistors — here the centre connection also formed the mechanical base for the structure. In modern practice, the base region is made as thin as possible to achieve reasonable levels of current gain; it is often only about one millionth of a meter thick.

Bipolar transistors are classified as either NPN or PNP according to the arrangement of their N-type and P-type materials. Their basic construction and chemical treatment is implied by their names. So an NPN transistors is formed by introducing a thin region of P-type material between two regions of N-type material.

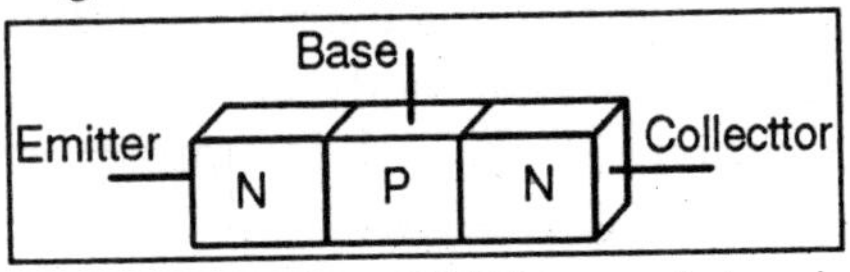


On the other hand, a PNP transistor is formed by introducing a thin region of N-type material between two regions of P-type material.

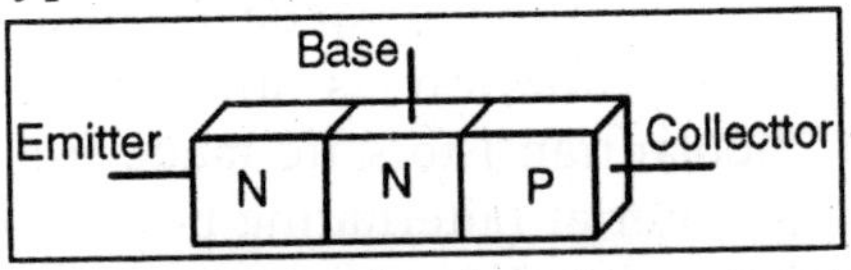


Since the majority and minority current carriers are different for N-type and P-type materials, it stands to reason that the internal operation of the NPN and PNP transistors will also be different. These two basic types of transistors along with their circuit symbols are shown here:

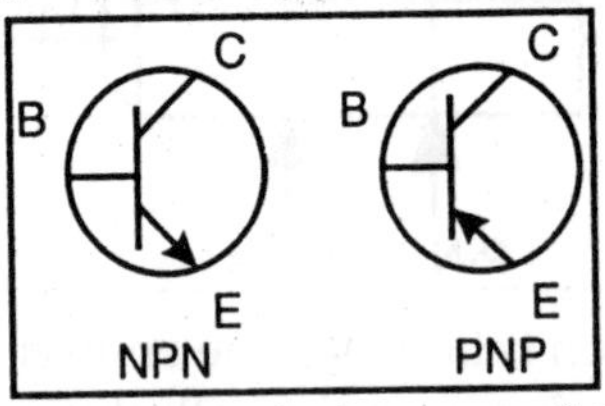


Note that the two symbols are subtly different. The vertical line represents the base (B), the angular line with the arrow on it represents the emitter (E), and the other angular line represents the collector (C). The direction of the arrow on the emitter distinguishes (graphically) the NPN from the PNP transistor. If the arrow points in, (Points iN) the transistor is a PNP. On the other hand if the arrow points out, the transistor is an NPN (Not Pointing iN).

Bear in mind that the arrow always points in the direction of hole flow (current), or from the P-type to N-type sections, no matter whether the P-type section is the emitter or base. On the other hand, electron flow is always "against" the arrow, just like in the junction diode.

As a result, a PNP transistor is "triggered" when its base is pulled low; an NPN transistor is "triggered" when its base is brought high.

Note that the bipolar transistor is a current-amplifying device, unlike the vacuum tube and the field-effect transistor (FET), both of which depend upon voltage changes to operate. It is the amount of current flowing in the base circuit that controls the amount of current flowing in the collector circuit.

## CONSTRUCTION

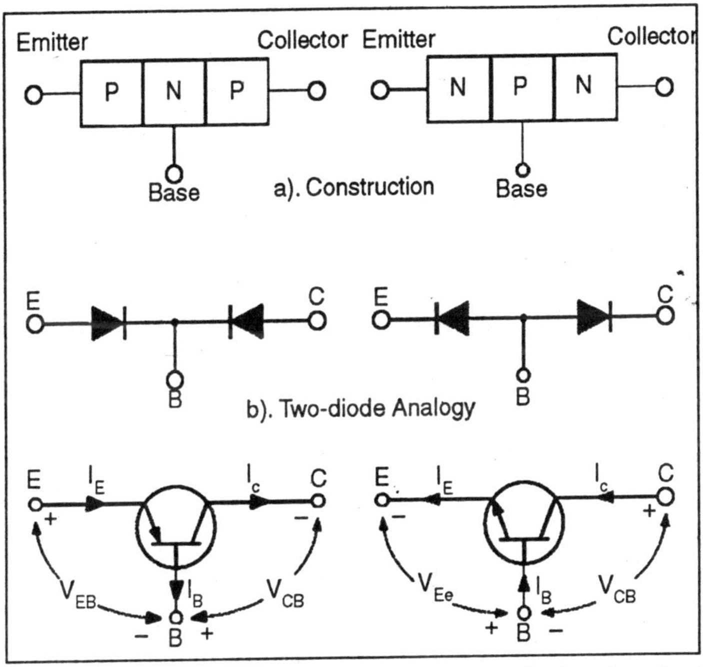


The construction and circuit symbols for both the NPN and PNP bipolar transistor are shown above with the arrow in the circuit symbol always showing the direction of conventional current flow between the base terminal and its emitter terminal, with the direction of the arrow pointing from the positive P-type region to the negative N-type region, exactly the same as for the standard diode symbol.

There are basically three possible ways to connect

a Bipolar Transistor within an electronic circuit with each method of connection responding differently to its input signal as the static characteristics of the transistor vary with each circuit arrangement.

- *Common Base Configuration*: Has Voltage Gain but no Current Gain.
- *Common Emitter Configuration*: Has both Current and Voltage Gain.
- *Common Collector Configuration*: Has Current Gain but no Voltage Gain.

## Common Base Configuration

As its name suggests, in the Common Base or Grounded Base configuration, the BASE connection is common to both the input signal AND the output signal with the input signal being applied between the base and the emitter terminals. The corresponding output signal is taken from between the base and the collector terminals as shown with the base terminal grounded or connected to a fixed reference voltage point. The input current flowing into the emitter is quite large as its the sum of both the base current and collector current respectively therefore, the collector current output is less than the emitter current input resulting in a Current Gain for this type of circuit of less than "1", or in other words it "Attenuates" the signal.

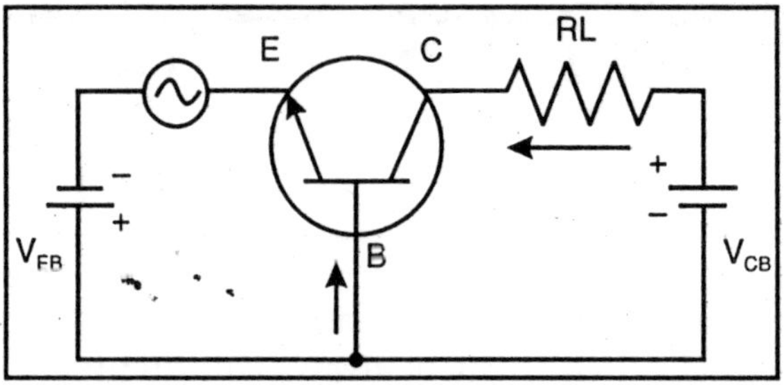


**Fig.** Common Base Amplifier Circuit

This type of amplifier configuration is a non-inverting voltage amplifier circuit, in that the signal voltages Vin and Vout are In-Phase. This type of arrangement is not very common due to its unusually high voltage gain characteristics.

Its Output characteristics represent that of a forward biased diode while the Input characteristics represent that of an illuminated photo-diode. Also this type of configuration has a high ratio of Output to Input resistance or more importantly "Load" resistance (RL) to "Input" resistance (Rin) giving it a value of "Resistance Gain". Then the Voltage Gain for a common base can therefore be given as:

*Common Base Voltage Gain*:

$$A_V = \frac{I_C \times R_L}{I_e \times R_{IN}} = \alpha \times \frac{R_L}{R_{IN}}$$

The Common Base circuit is generally only used in single stage amplifier circuits such as microphone pre-amplifier or RF radio amplifiers due to its very good high frequency response.

## Common Emitter Configuration

In the Common Emitter or Grounded Emitter configur-ation, the input signal is applied between the base, while the output is taken from between the collector and the emitter as shown. This type of configuration is the most commonly used circuit for transistor based amplifiers and which represents the "normal" method of connection. The common emitter amplifier configuration produces the highest current and power gain of all the three bipolar transistor configurations. This is mainly because the input impedance is LOW as it is connected to a forward-biased junction, while the output impedance is HIGH as it is taken from a reverse-biased junction.

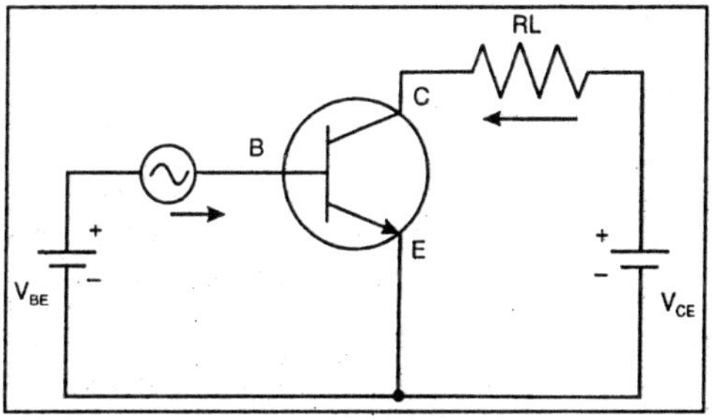


**Fig.** Common Emitter Amplifier Circuit

In this type of configuration, the current flowing out of the transistor must be equal to the currents flowing into the transistor as the emitter current is given as Ie = Ic + Ib. Also, as the load resistance (RL) is connected in series with the collector, the Current gain of the Common Emitter Transistor Amplifier is quite large as it is the ratio of Ic/Ib and is given the symbol of Beta, (β). Since the relationship between these three currents is determined by the transistor itself, any small change in the base current will result in a large change in the collector current. Then, small changes in base current will thus control the current in the Emitter/Collector circuit.

By combining the expressions for both Alpha, α and Beta, β the mathematical relationship between these parameters and therefore the current gain of the amplifier can be given as:

$$I_E = I_C + I_B$$

$$\alpha = \frac{I_C}{I_C} \, and \, \beta = \frac{I_C}{I_B}$$

$$\alpha = \frac{\beta}{\beta + 1} \quad \beta = \frac{\alpha}{1 - \alpha}$$

Where: "Ic" is the current flowing into the collector terminal, "Ib" is the current flowing into the base terminal and "Ie" is the current flowing out of the emitter terminal.

Then to summarise, this type of bipolar transistor configuration has a greater input impedance, Current and Power gain than that of the common Base configuration but its Voltage gain is much lower. The common emitter is an inverting amplifier circuit resulting in the output signal being 180° out of phase with the input voltage signal.

### Common Collector Configuration

In the Common Collector or Grounded Collector configuration, the collector is now common and the input signal is connected to the Base, while the output is taken from the Emitter load as shown. This type of configuration is commonly known as a Voltage Follower or Emitter Follower circuit. The Emitter follower configuration is very useful for impedance matching applications because of the very high input impedance, in the region of hundreds of thousands of Ohms, and it has relatively low output impedance.

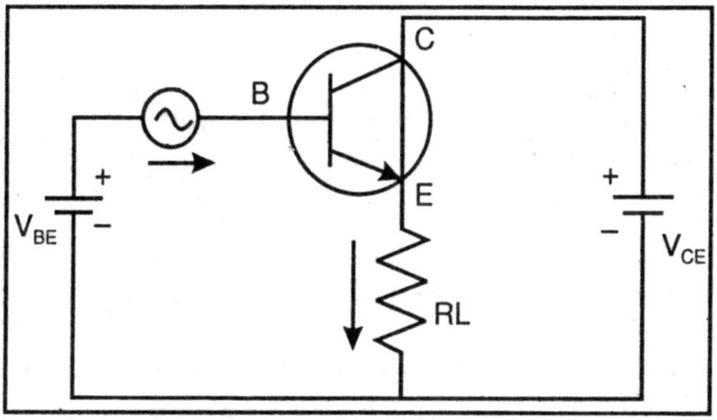


**Fig.** Common Collector Amplifier Circuit

The Common Emitter configuration has a current gain equal to the β value of the transistor itself. In the common collector configuration the load resistance is situated in series with the emitter so its current is equal to that of the emitter current.

As the emitter current is the combination of the collector AND base currents combined, the load resistance in this type of amplifier configuration also has both the collector current and the input current of the base flowing through it.

*Then the current gain of the circuit is given as:*

$$I_E = I_C + I_B$$

$$A_i = \frac{I_E}{I_B} = \frac{I_C + I_B}{I_B}$$

$$A_i = \frac{I_C}{I_B} + 1$$

$$\therefore A_i = \beta + 1$$

This type of bipolar transistor configuration is a non-inverting amplifier circuit in that the signal voltages of Vin and Vout are "In-Phase". It has a voltage gain that is always less than "1" (unity). The load resistance of the common collector amplifier configuration receives both the base and collector currents giving a large current gain (as with the

Common Emitter configuration) therefore, providing good current amplification with very little voltage gain.

## FUNDAMENTALS OF BJT OPERATION

A bipolar junction transistor consists of two back-to-back p-n junctions, who share a thin common region with width, $w_B$. Contacts are made to all three regions, the two outer regions called the emitter and collector and the middle region called the base.

The structure of an npn bipolar transistor is shown in Figure below (a). The device is called "bipolar" since its operation involves both types of mobile carriers, electrons and holes.

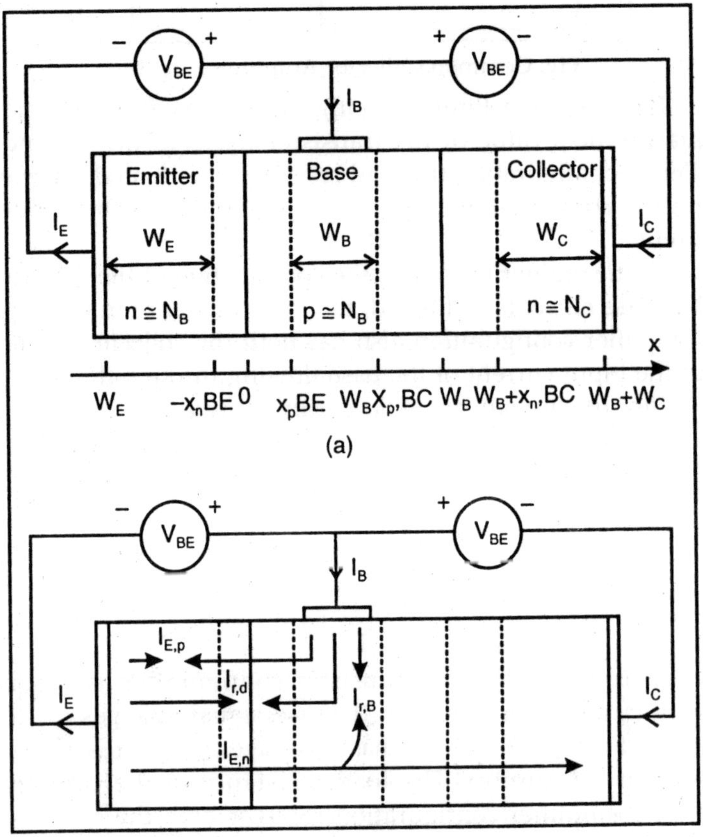

(a) Structure and sign convention of a npn bipolar junction transistor.

(b) Electron and hole flow under forward active bias, $V_{BE} > 0$ and $V_{BC} = 0$.

Since the device consists of two back-to-back diodes, there are depletion regions between the quasi-neutral regions. The width of the quasi neutral regions in the emitter, base and collector are indicated with the symbols $w_E'$, $w_B'$ and $w_C'$ and are calculated from,

$$w_E' = w_E - x_n, BE$$
$$w_B' = w_B - x_p, BE^{-x}p, BC$$
$$w_C' = w_C - x_n, BC$$

where the depletion region widths are given by:

$$x_n, BE = \sqrt{\frac{2\varepsilon_s(\varphi_i, BE - V_{BE})}{q}\frac{N_B}{N_E}\left(\frac{1}{N_B + N_E}\right)}$$

$$x_p, BE = \sqrt{\frac{2\varepsilon_s(\varphi_i, BE - V_{BE})}{q}\frac{N_E}{N_B}\left(\frac{1}{N_B + N_E}\right)}$$

$$x_p, BC = \sqrt{\frac{2\varepsilon_s(\varphi_i, BE - V_{BC})}{q}\frac{N_C}{N_B}\left(\frac{1}{N_B + N_E}\right)}$$

$$x_n, BC = \sqrt{\frac{2\varepsilon_s(\varphi_i, BC - V_{BC})}{q}\frac{N_B}{N_C}\left(\frac{1}{N_B + N_E}\right)}$$

with

$$\varphi_i, BE = V_t \, In\frac{N_E N_B}{n_i^2}$$

$$\varphi_i, BC = V_t \, In\frac{N_C N_B}{n_i^2}$$

The sign convention of the currents and voltage is indicated on Figure above in (a). The base and collector current are positive if a positive current goes into the base or collector contact. The emitter current is positive for a current coming out of the emitter contact. This also implies that the emitter current, $I_E$, equals the sum of the base current, $I_B$, and the collector current, $I_C$:

$$I_E = I_C + I_B$$

The base-emitter voltage and the base-collector voltage are positive if a positive voltage is applied to the base contact relative to the emitter and collector respectively.

The operation of the device is illustrated with Figure above (b). We consider here only the forward active bias mode of operation, obtained by forward biasing the base-emitter junction and reverse biasing the base-collector junction. To simplify the discussion further, we also set $V_{CE} = 0$. The corresponding energy band diagram is shown in Figure below. Electrons diffuse from the emitter into the base and holes diffuse from the base into the emitter. This carrier diffusion is identical to that in a p-n junction.

However, what is different is that the electrons can diffuse as minority carriers through the quasi-neutral base. Once the electrons arrive at the base-collector depletion region, they are swept through the depletion layer due to the electric field.

These electrons contribute to the collector current. In addition, there are two more currents, the base recombination current, indicated on Figure below by the vertical arrow, and the base-emitter depletion layer recombination current, $I_{r,d}$, (not shown).

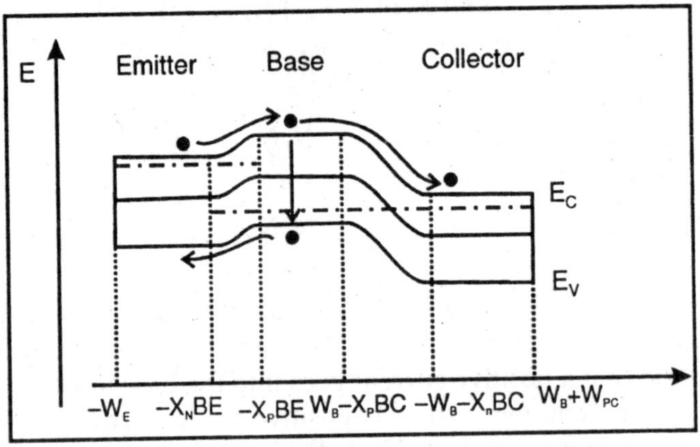


**Fig.** Energy Band Diagram of a Bipolar Transistor Biased in the Forward Active Mode.

The total emitter current is the sum of the electron

diffusion current, $I_{E,n}$, the hole diffusion current, $I_{E,p}$ and the base-emitter depletion layer recom-bination current, $I_{r,d}$.

$$I_E = I_{E,n} + I_{E,P} + I_{r,d}$$

The total collector current is the electron diffusion current, $I_{E,n}$, minus the base recombination current, $I_{r,B}$.

$$I_C = I_{E,n} - I_{r,B}$$

The base current is the sum of the hole diffusion current, $I_{E,p}$, the base recombination current, $I_{r,B}$ and the base-emitter depletion layer recombination current, $I_{r,d}$.

$$I_B = I_{E,P} + I_{r,B} + I_{r,d}$$

The transport factor, α, is defined as the ratio of the collector and emitter current

$$\alpha = \frac{I_C}{I_E}$$

Using Kirchoff's current law and the sign convention shown in Figure above in (a), we find that the base current equals the difference between the emitter and collector current. The current gain, β, is defined as the ratio of the collector and base current and equals:

$$\beta = \frac{I_C}{I_E} = \frac{\alpha}{1-\alpha}$$

This explains how a bipolar junction transistor can provide current amplification. If the collector current is almost equal to the emitter current, the transport factor, a, approaches one.

The current gain, β, can therefore become much larger than one. To facilitate further analysis, we now rewrite the transport factor, α, as the product of the emitter efficiency, $\gamma_E$, the base transport factor, $\alpha_T$, and the depletion layer recombination factor, $\delta_r$.

$$\alpha = \alpha_T \gamma_E \delta_r$$

The emitter efficiency, $g_E$, is defined as the ratio of the electron current in the emitter, $I_{E,n}$, to the sum of the electron and hole current diffusing across the base-emitter junction, $I_{E,n} + I_{E,p}$.

$$\gamma_E = \frac{I_{E.n}}{I_{E,N} + I_{E,p}}$$

The base transport factor, $a_T$, equals the ratio of the current due to electrons injected in the collector, to the current due to electrons injected in the base.

$$\alpha_T = \frac{I_{E.n} - I_{r,B}}{I_{E},n}$$

Recombination in the depletion-region of the base-emitter junction further reduces the current gain, as it increases the emitter current without increasing the collector current. The depletion layer recombination factor, $d_r$, equals the ratio of the current due to electron and hole diffusion across the base-emitter junction to the total emitter current:

$$\delta_r = \frac{I_E - I_{r,d}}{I_E}$$

Example: A bipolar transistor with an emitter current of 1 mA has an emitter efficiency of 0.99, a base transport factor of 0.995 and a depletion layer recombination factor of 0.998. Calculate the base current, the collector current, the transport factor and the current gain of the transistor.

Solution

The transport factor and current gain are:

$$\alpha = \gamma E \alpha T \delta_\gamma = 0.99 \times 0.995 \times 0.998 = 0.9983$$

and

$$\beta = \frac{\alpha}{1-\alpha} = 58.1$$

The collector current then equals

$$I_C = \alpha I_E = 0.983 mA$$

And the base current is obtained from:

$$I_B = I_E - I_C = 17 \mu A$$

## AMPLIFICATION WITH BJTS

### Amplification Classes (III)

When communicating with another ham the main objective is to emit a signal displaying an excellent modulation that please to your correspondent's ears. Single-sideband transmitters generate the carrier and modulation at a low frequency and translate it up with a mixer. Linear amplifiers are used to increase the power level up to say 100W. If we have to amplify this signal to higher values because the station if far away, we want that our amplified signal be strictly identical in shape to the original, only its intensity must have increased. We call this characteristics, the linearity.

### *Linearity*

Several factors determine the linearity of an HF so-called "linear" amplifier:

- Changes in device power gain (ratio of output to drive power) over the range from zero to full output
- The collector/emitter (BJT transistor) or drain/source (MOSFET transistor) peak RF voltage excursion
- The regulation (stiffness) of the collector (+Vcc) or drain (+Vdd) supply voltage
- The standing (idle) current and the onset of saturation in the RF transformers.

The limiting factor is the constancy of power gain over the entire power-output excursion. To visualise this, one can inspect the curve of output vs. drive power in the transistor data sheet. Generally, MOSFETs will exhibit superior linearity as compared to bipolar junction transistors (BJTs). MOSFETs such as the Motorola MRF150 began to displace BJTs (e.g. MRF448, 2SC2652) in the late 1990's.

The higher the voltage excursion, the longer the linear portion of the output/drive power curve. For this reason, amplifiers powered between 40-50V DC exhibit considerably better linearity, and thus lower IMD, than 13.8V units. Typical 3$^{rd}$-order IMD (IMD3) values relative to PEP are –32 dB for 50V and –24 dB for 13.8V. Additionally, it is more difficult to design a power supply for a 13.8 V amplifier due to high current requirements (e.g. typically 80A peak for a 500W PEP amplifier like Ameritron ALS-500MX). So usually a tube amplifier is more linear than any solid-state model (typically IMD3 < –40 dB relative to PEP).

The collector-to-collector (or drain-to-drain) load impedance must be maintained as close to resistive as possible. This in turn requires optimising the wideband output transformer(s), combiner (if used), and low-pass filter passband VSWRs to as low values as possible. The load presented to the low-pass filter output must also be as close as possible to 50 ohms resistive.

The amplification class defines the functioning mode of the electronic circuit. It has been divided in several classes: A, B, AB, AB1, AB2, C, D, E, G, H. Class A is the most appreciated by fans of high fidelity audio and amateur radio because this is the mode in which the signal yield the best performances, in both musicality (for audio applications) and output. However this class has some drawbacks and therefore hams prefer using power devices in Class AB, so many classes that we are going to review.

*Class A*

To work in Class A, typically the signal to amplify has a working point (quiescent point) that normally lies in the middle of the active region of the load line, as displayed at left.

At rest, without input signal, the cathode current flows contineously in the device (tube or transistor), consuming at lot of power, proportional to $V_{ce}I_c$. One says that this mode offers a conducting angle of 360°.

To avoid distorsions (saturation or blocking) the peak values for the output current and voltage must be always smaller than the working point voltage and current. The maximum signal power of this device is limited by the relationships:

$$P_{max} = \frac{U_{CE}I_C}{2} P_{dc} = U_b I_c$$

In Class A the output signal is directly proportional to the input; hence the name of linear amplifier. This mode is ideal if you are looking for high quality signals but it is paid by the consumption of much energy. Its efficiency is at best 50%.

That means that more than a half of the power is lost in heat, and indeed these amplifiers heat much and request an oversized radiator to dissipate the lost power...

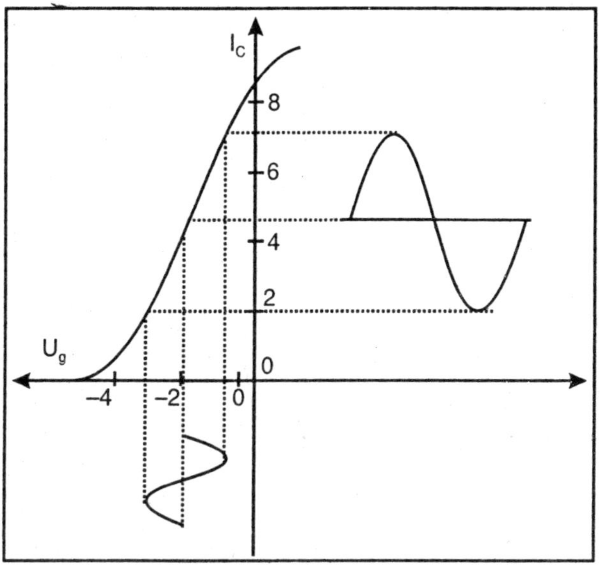


Class A is the widely used amplification mode in low frequencies and in the first HF stages of radio equipments. And as you probably know if you have a Hi-Fi system, in audio it is the privileged amplification class for the purists.

*Class B*

Class A or Class B amplifiers are used to increase the power level of weak AM or SSB signals. But Class A amplification yield a low output due to much power dissipation. The best should be to suppress the polarization current in quiescent mode.

This is achieved in Class B amplifiers that are biased at cutoff (blocking point) so that the collector current flows for 180° of the input cycle. In other words, as soon as there is an input signal, the device unlocks during half a cycle and conducts.

If we suspect that this tube or transistor will yield a better output as it does no more consume power at rest (quiescent mode), the output signal does only return one alternation of two and we will guess that this mode is not satisfying. Then, as the device will only work in the presence of an input signal, for weak inputs it will not work in the linear region of its characteristics and thus it will not be linear.

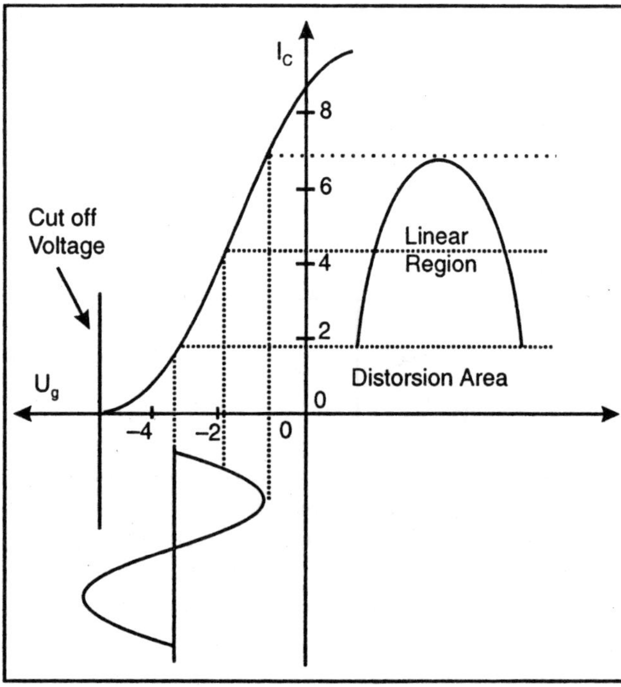


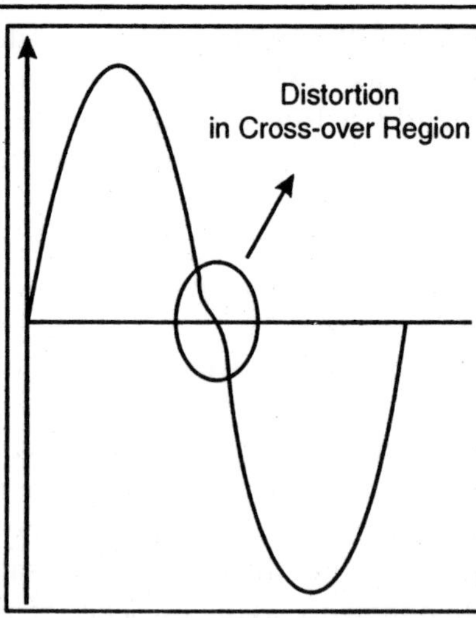


At left when a Class B amplifier works close to cut off, its transistors or tubes can only amplify one half of the signal. Therefore class B amplifiers will always consist of two transistors or tubes (push-pull circuit), where each component will supply one half wave of the output signal. At right the control characteristics of two transistors of a Class B amplifier. The non-linearity of the characteristic near zero-current produces cross-over distortions. One correct this effect with Class AB.

To amplify the same way both alternations, Class B amplifiers are usually connected in a push-pull circuit where one transistor amplifies each half of the input signal; there are thus 2 transitors or tubes, each of them amplifying either the positive or the negative alternation.

To create the opposite wave for each component, a phase splitter or phase inverter task is inserted in the circuit. But here also there is a problem in the cross-over region that creates a distortion at the time of alternation is changing.

One correct this effect in increasing the conducting angle so that the phase is greater than 180°. This lead to the Class AB amplification. In summary, Class B yield a good output with an efficiency of 60%. Drawbacks are the cross-over distortion, unsuited for amateur emissions and the necessity to install transistors or tubes controlled in anti-phase to get a symmetrical set.

*Class AB (IV)*

In Class A we have discovered that the ouput power is quite limited and we cannot reach high signals while in Class B the ouput signal is more powerful but yield no linearity, and both classes show a too low efficiency.

As amateur radio we need power, 100W in input and up to 2 kW ouput (that depends of your national regulation), we need a linear signal and efficiency. To get these results, electronicians invented the Class AB and its derivated Classes AB1 and AB2. Today all linear amplifiers used by hams belong to Class AB which efficiency does not exceed 50%, there are tubes or solid-state amplifiers. These amplifiers use power devices that are are slightly polarized so that they work in the linear region of the load line for weak signals (Class A) and in Class B for stronger signals. In this way we avoid cross-over distortions as the working point of transistors or tubes can be placed a bit higher than the cut off at the lower end of the active region.

This polarization is less than in Class A, allowing the transistor or tube to emit very few current when there is no signal in input. But as they are somewhat, the linearity is modified. Therefore in all amateurs amplifiers the output circuit is completed with a filtering function to cut all harmonics produced in this process.

Class AB is subdivided in Class AB1 and AB2 depending the power stage characteristics Class AB1 concerns a power stage that does not absorb current from the previous stage; this is usually the case with vacuum tubes without grid current and the Field Effect Transistors, the famous FETs.

This stage yield usually a high impedance Class AB2 concerns a power stage that absorbs some current from the previous stage; some vaccuum tubes are used in this manner and all the classic bipolar transistors.

In the field it is easy to recognize a ham using a Class AB amplifier so-called linear. When speaking some of his sentences are grinded and works lack in the conversation. This effect occurs when the polarization is too low, with for consequence to transmit only the peaks of modulation; the current is transmitted during less than half of a wave of the output signal. To solve this problem this ham has to increase the quiescent current.

REVIEW:

- In digital circuits, binary bit values of 0 and 1 are represented by voltage signals measured in reference to a common circuit point called ground. An absence of voltage represents a binary "0" and the presence of full DC supply voltage represents a binary "1."
- A logic gate, or simply gate, is a special form of amplifier circuit designed to input and output logic level voltages (voltages intended to represent binary bits). Gate circuits are most commonly represented

in a schematic by their own unique symbols rather than by their constituent transistors and resistors.

- Just as with operational amplifiers, the power supply connections to gates are often omitted in schematic diagrams for the sake of simplicity.
- A truth table is a standard way of representing the input/output relationships of a gate circuit, listing all the possible input logic level combinations with their respective output logic levels.

## Chapter 2

# Boolean Algebra

Mathematical rules are based on the defining limits we place on the particular numerical quantities dealt with. When we say that,

$$1 + 1 = 2$$

or

$$3 + 4 = 7$$

we are implying the use of integer quantities: the same types of numbers we all learned to count in elementary education. What most people assume to be self-evident rules of arithmetic valid at all times and for all purposes — actually depend on what we define a number to be.

For instance, when calculating quantities in AC circuits, we find that the "real" number quantities which served us so well in DC circuit analysis are inadequate for the task of representing AC quantities. We know that voltages add when connected in series, but we also know that it is possible to connect a 3-volt AC source in series with a 4-volt AC source and end up with 5 volts total voltage (3 + 4 = 5)! Does this mean the inviolable and self-evident rules of arithmetic have been violated?

No, it just means that the rules of "real" numbers do not apply to the kinds of quantities encountered in AC circuits, where every variable has both a magnitude and a phase. Consequently, we must use a different kind of numerical quantity, or object, for AC circuits (*complex* numbers, rather than *real* numbers), and along with this different system of numbers comes a different set of rules telling us how they relate to one another.

An expression such as "3 + 4 = 5" is nonsense within the scope and definition of real numbers, but it fits nicely within the scope and definition of complex numbers (think of a right triangle with opposite and adjacent sides of 3 and 4, with a hypotenuse of 5). Because complex numbers are two-dimensional, they are able to "add" with one another trigonometrically as single-dimension "real" numbers cannot.

Logic is much like mathematics in this respect: the so-called "Laws" of logic depend on how we define what a proposition is. The Greek philosopher Aristotle founded a system of logic based on only two types of propositions: true and false. His bivalent (two-mode) definition of truth led to the four foundational laws of logic: the Law of Identity (A is A); the Law of Non-contradiction (A is not non-A); the Law of the Excluded Middle (either A or non-A); and the Law of Rational Inference.

These so-called Laws function within the scope of logic where a proposition is limited to one of two possible values, but may not apply in cases where propositions can hold values other than "true" or "false." In fact, much work has been done and continues to be done on "multivalued," or *fuzzy* logic, where propositions may be true or false *to a limited degree.*

In such a system of logic, "Laws" such as the Law of the Excluded Middle simply do not apply, because they are founded on the assumption of bivalence. Likewise, many premises which would violate the Law of Non-contradiction in Aristotelian logic have validity in "fuzzy" logic. Again, the defining limits of propositional values determine the Laws describing their functions and relations.

Boole wrote a treatise on the subject in 1854, titled *An Investigation of the Laws of Thought, on Which Are Founded the Mathematical Theories of Logic and Probabilities,* which codified several rules of relationship between mathematical quantities limited to one of two possible values: true or false, 1 or 0. His mathematical system became known as Boolean algebra.

All arithmetic operations performed with Boolean quantities have but one of two possible outcomes: either 1 or 0. There is no such thing as "2" or "-1" or "1/2" in the Boolean world. It is a world in which all other possibilities are invalid by fiat.

As one might guess, this is not the kind of math you want to use when balancing a checkbook or calculating current through a resistor. However, Claude Shannon of MIT fame recognized how Boolean algebra could be applied to on-and-off circuits, where all signals are characterized as either "high" (1) or "low" (0).

The thesis, titled *A Symbolic Analysis of Relay and Switching Circuits*, put Boole's theoretical work to use in a way Boole never could have imagined, giving us a powerful mathematical tool for designing and analyzing digital circuits.

In this chapter, you will find a lot of similarities between Boolean algebra and "normal" algebra, the kind of algebra involving so-called real numbers. Just bear in mind that the system of numbers defining Boolean algebra is severely limited in terms of scope, and that there can only be one of two possible values for any Boolean variable: 1 or 0.

Consequently, the "Laws" of Boolean algebra often differ from the "Laws" of real-number algebra, making possible such statements as 1 + 1 = 1, which would normally be considered absurd. Once you comprehend the premise of all quantities in Boolean algebra being limited to the two possibilities of 1 and 0, and the general philosophical principle of Laws depending on quantitative definitions, the "nonsense" of Boolean algebra disappears.It should be clearly understood that Boolean numbers are not the same as *binary* numbers.

Whereas Boolean numbers represent an entirely different system of mathematics from real numbers, binary is nothing more than an alternative *notation* for real numbers. The two are often confused because both Boolean math and binary notation use the same two ciphers: 1 and 0.The difference is that Boolean quantities are restricted to a single bit (either 1 or 0), whereas binary numbers may be composed of many bits adding up in place-weighted form to a value of any finite size. The binary number $10011_2$("nineteen") has no more place in the Boolean world than the decimal number $2_{10}$ ("two") or the octal number $32_8$ ("twenty-six").

## BOOLEAN ARITHMETIC

In computer science, the Boolean or logical data type is a primitive data type having one of two values: true or false, intended to represent the truth values of logic and Boolean algebra. In programming languages that have a built-in Boolean data type, such as Pascal and Java, the comparison operators such as'>' and'?' are usually defined to return a Boolean value. Also, conditional and iterative commands may be defined to test Boolean-valued expressions.

Languages without an explicit Boolean data type, like C and Lisp, may still represent truth values it by some other data type. Lisp uses an empty list for false, and any other value for true. C uses an integer type, with false represented as the zero value, and true as any non-zero value (such as 1 or -1). Indeed, a Boolean variable may be regarded (and be implemented) as a numerical variable with a single binary digit (bit), which can store only two values. Let us begin our exploration of Boolean algebra by adding numbers together:

$$0 + 0 = 0$$
$$0 + 1 = 1$$
$$1 + 0 = 1$$
$$1 + 1 = 1$$

The first three sums make perfect sense to anyone familiar with elementary addition. The last sum, though, is quite possibly responsible for more confusion than any other single statement in digital electronics, because it seems to run contrary to the basic principles of mathematics. Well, it *does* contradict principles of addition for real numbers, but not of elimination.

It does not matter how many or few terms we add together, either.

*Consider the following sums:*

$0 + 1 + 1 = 1$

$1 + 1 + 1 = 1$

$0 + 1 + 1 + 1 = 1$

$1 + 0 + 1 + 1 + 1 = 1$

Take a close look at the two-term sums in the first set of equations. Does that pattern look familiar to you? It should! It is the same pattern of 1's and 0's as seen in the truth table for an OR gate. In other words, Boolean addition corresponds to the logical function of an "OR" gate, as well as to parallel switch contacts:

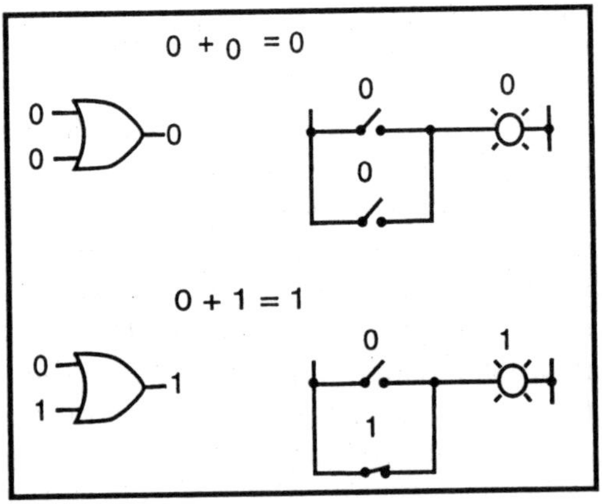


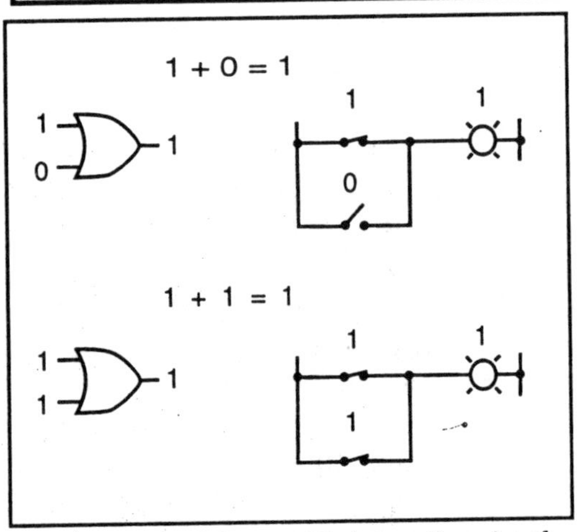


There is no such thing as subtraction in the realm of Boolean mathematics. Subtraction implies the existence of negative numbers: 5 – 3 is the same thing as 5 + (–3), and in Boolean algebra negative quantities are forbidden.

There is no such thing as division in Boolean mathematics, either, since division is really nothing more than compounded subtraction, in the same way that multiplication is compounded addition.

Multiplication is valid in Boolean algebra, and thankfully it is the same as in real-number algebra: anything multiplied by 0 is 0, and anything multiplied by 1 remains unchanged:

$$0 \times 0 = 0$$
$$0 \times 1 = 0$$
$$1 \times 0 = 0$$
$$1 \times 1 = 1$$

This set of equations should also look familiar to you: it is the same pattern found in the truth table for an AND gate. In other words, Boolean multiplication corresponds to the logical function of an "AND" gate, as well as to series switch contacts:

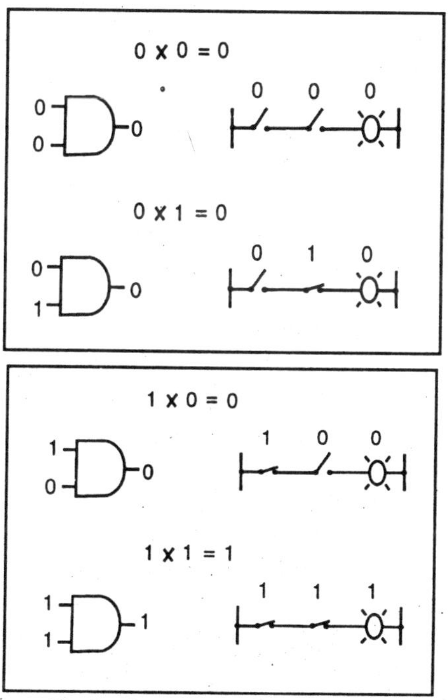


Like "normal" algebra, Boolean algebra uses alphabetical letters to denote variables. Unlike "normal" algebra, though, Boolean variables are always CAPITAL letters, never lower-case. Because they are allowed to possess only one of two possible values, either 1 or 0, each and every variable has a*complement*: the opposite of its value.

For example, if variable "A" has a value of 0, then the complement of A has a value of 1. Boolean notation uses a bar above the variable character to denote complementation, like this:

*If*: $A = 0$

*Then*: $\overline{A} = 1$

*If*: $A = 1$

*Then*: $\overline{A} = 1$

In written form, the complement of "A" denoted as "A-not" or "A-bar". Sometimes a "prime" symbol is used to represent complementation. For example, A′ would be the complement of A, much the same as using a prime symbol to denote differentiation in calculus rather than the fractional notation d/dt.

Usually, though, the "bar" symbol finds more widespread use than the "prime" symbol, for reasons that will become more apparent later in this chapter.

Boolean complementation finds equivalency in the form of the NOT gate, or a normally-closed switch or relay contact:

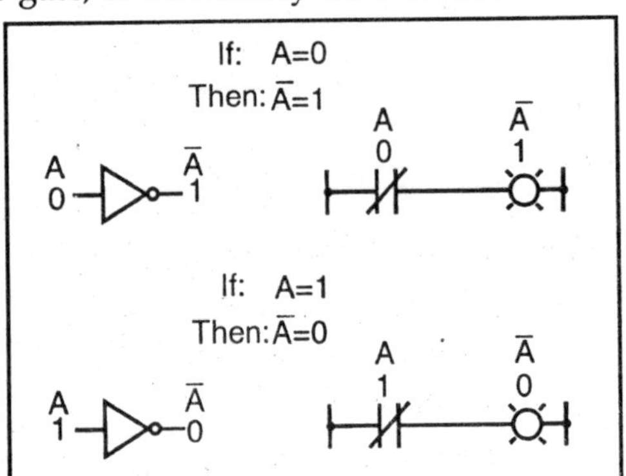


The basic definition of Boolean quantities has led to the simple rules of addition and multiplication, and has excluded both subtraction and division as valid arithmetic operations. We have a symbology for denoting Boolean variables, and their complements. In the next section we will proceed to develop Boolean identities.

- *Review:*
- Boolean addition is equivalent to the *OR* logic function, as well as *parallel* switch contacts.
- Boolean multiplication is equivalent to the *AND* logic function, as well as *series* switch contacts.
- Boolean complementation is equivalent to the *NOT* logic function, as well as *normally-closed*relay contacts.

## BOOLEAN ALGEBRAIC IDENTITIES

In mathematics, an *identity* is a statement true for all possible values of its variable or variables. The algebraic identity of x + 0 = x tells us that anything (x) added to zero equals the original "anything," no matter what value that "anything" (x) may be. Like ordinary algebra, Boolean algebra has its own unique identities based on the bivalent states of Boolean variables.

The first Boolean identity is that the sum of anything and zero is the same as the original "anything." This identity is no different from its real-number algebraic equivalent:

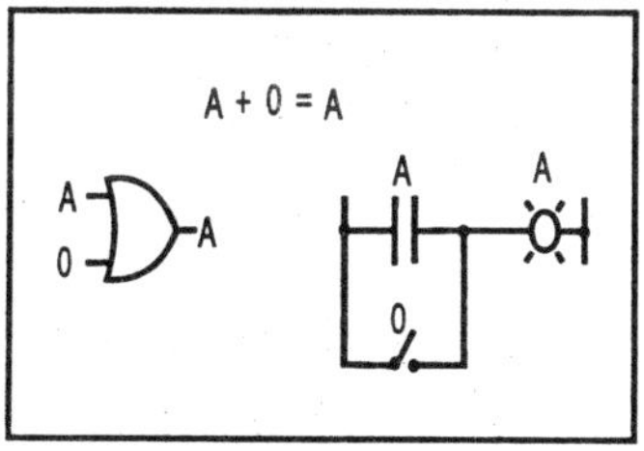


No matter what the value of A, the output will always be the same: when A=1, the output will also be 1; when A=0, the output will also be 0. The next identity is most definitely *different* from any seen in normal algebra. Here we discover that the sum of anything and one is one:

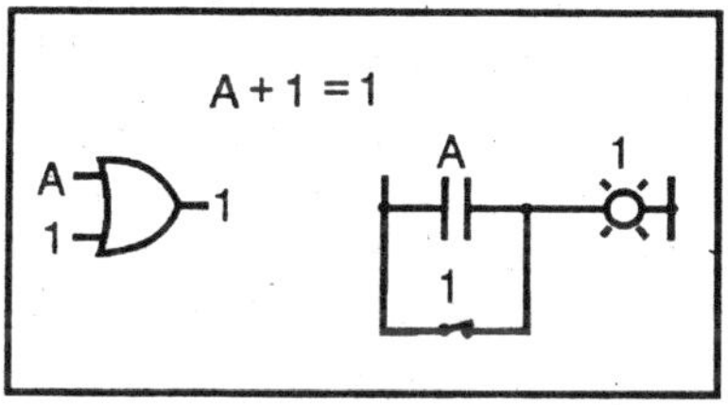

No matter what the value of A, the sum of A and 1 will always be 1. In a sense, the "1" signal *overrides*the effect of A on the logic circuit, leaving the output fixed at a logic level of 1.Next, we examine the effect of adding A and A together, which is the same as connecting both inputs of an OR gate to each other and activating them with the same signal:

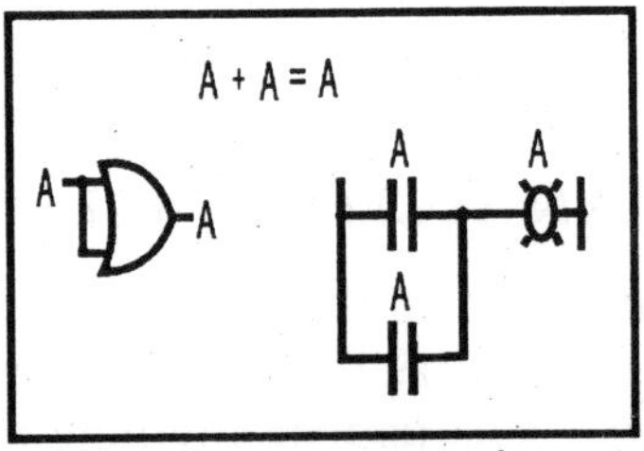


In real-number algebra, the sum of two identical variables is twice the original variable's value (x + x = 2x), but remember that there is no concept of "2" in the world of Boolean math, only 1 and 0, so we cannot say that A + A = 2A. Thus, when we add a Boolean quantity to itself, the sum is equal to the original quantity:

$$0 + 0 = 0,$$
and
$$1 + 1 = 1.$$

Introducing the uniquely Boolean concept of complementation into an additive identity, we find an interesting effect. Since there must be one "1" value between any variable and its complement, and since the sum of any Boolean quantity and 1 is 1, the sum of a variable and its complement must be 1:

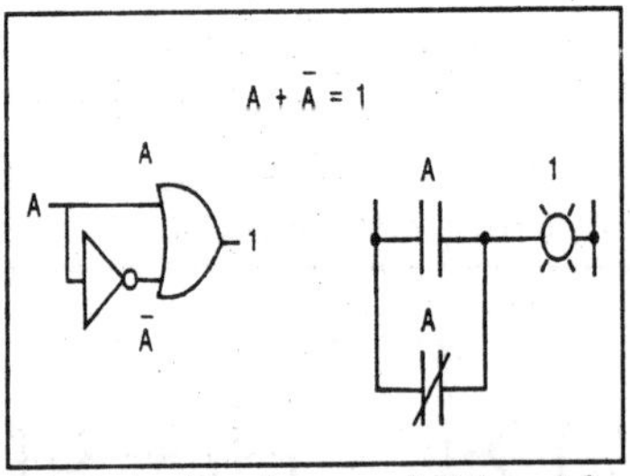


Just as there are four Boolean additive identities (A+0, A+1, A+A, and A+A'), so there are also four multiplicative identities:

A × 0, A × 1, A × A, and A × A′. Of these, the first two are no different from their equivalent expressions in regular algebra:

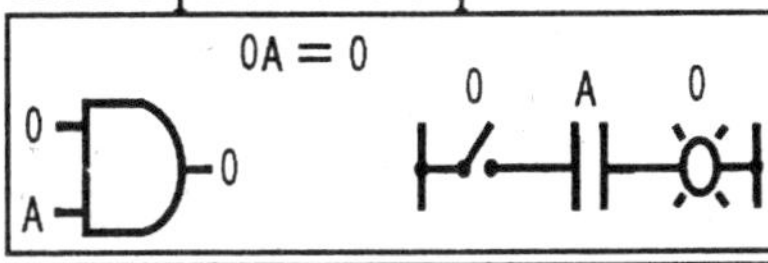


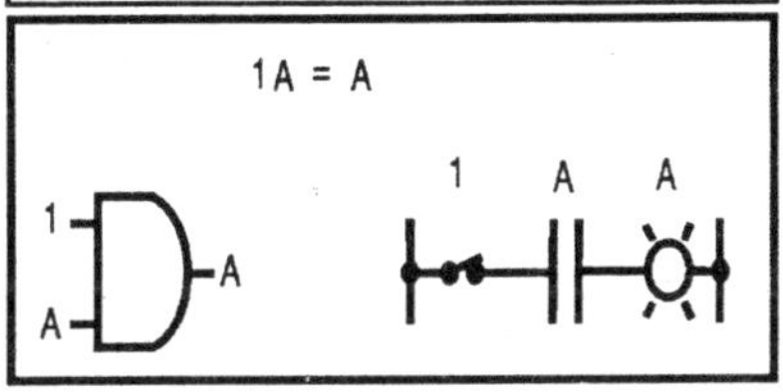


The third multiplicative identity expresses the result of a Boolean quantity multiplied by itself. In normal algebra, the product of a variable and itself is the *square* of that variable (3 × 3 = $3^2$ = 9). However, the concept of "square" implies a quantity of 2, which has no meaning in Boolean algebra, so we cannot say that A × A = $A^2$. Instead, we find that the product of a Boolean quantity and itself is the original quantity, since 0 × 0 = 0 and 1 × 1 = 1:

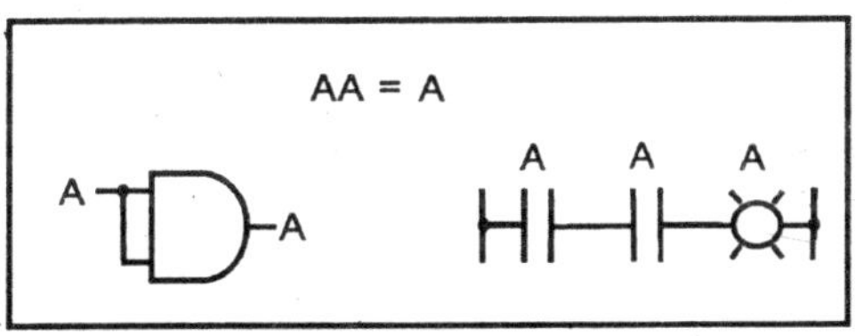


The fourth multiplicative identity has no equivalent in regular algebra because it uses the complement of a variable, a concept unique to Boolean mathematics. Since there must be one "0" value between any variable and its complement, and since the product of any Boolean quantity and 0 is 0, the product of a variable and its complement must be 0:

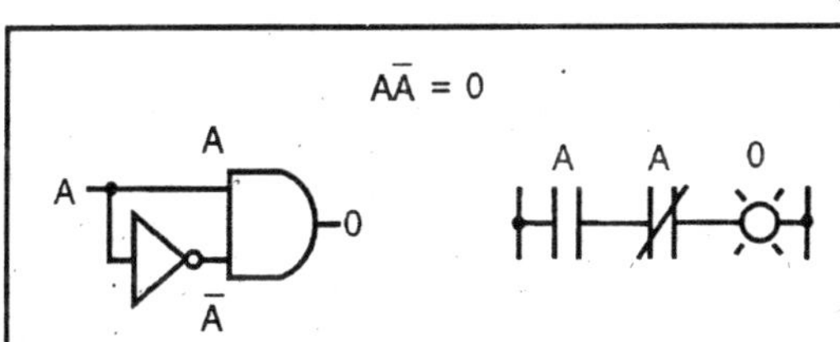

To summarize, then, we have four basic Boolean identities for addition and four for multiplication:

**Table. Basic Boolean Algebraic Identities**

| Additive | Multiplicative |
|---|---|
| $A + 0 = A$ | $0A = 0$ |
| $A + 1 = 1$ | $1A = A$ |
| $A + A = A$ | $AA = A$ |
| $A + \overline{A} = 1$ | $A\overline{A} = 0$ |

Another identity having to do with complementation is that of the *double complement*: a variable inverted twice. Complementing a variable twice (or any even number of times) results in the original Boolean value. This is analogous to negating (multiplying by -1) in real-number algebra: an even number of negations cancel to leave the original value:

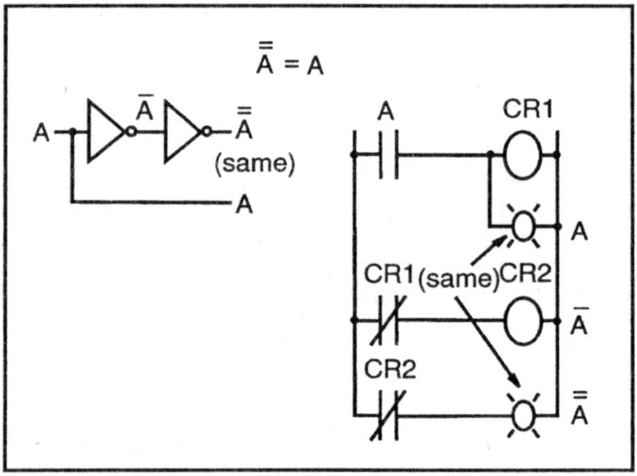


## BOOLEAN ALGEBRA PROPERTIES

Another type of mathematical identity, called a "property" or a "law," describes how differing variables relate to each other in a system of numbers. One of these properties is known as the *commutative property,* and it applies equally to addition and multiplication. In essence, the commutative property tells us we can reverse the order of variables that are either added together or multiplied together without changing the truth of the expression:

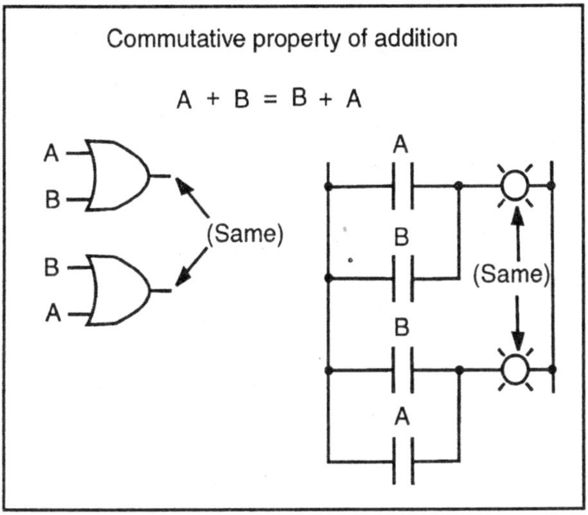


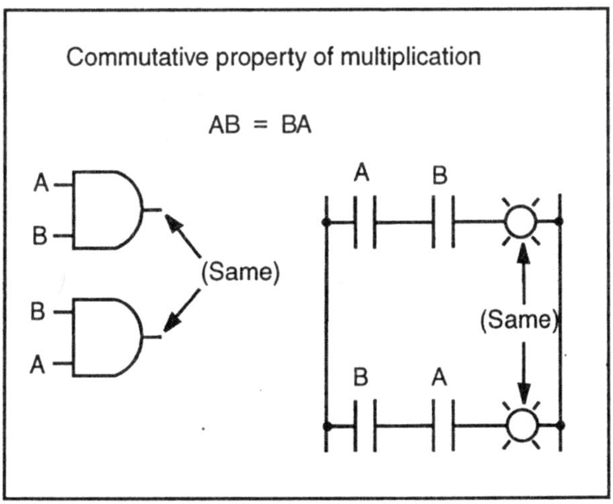


Along with the commutative properties of addition and multiplication, we have the *associative property*, again applying equally well to addition and multiplication. This property tells us we can associate groups of added or multiplied variables together with parentheses without altering the truth of the equations.

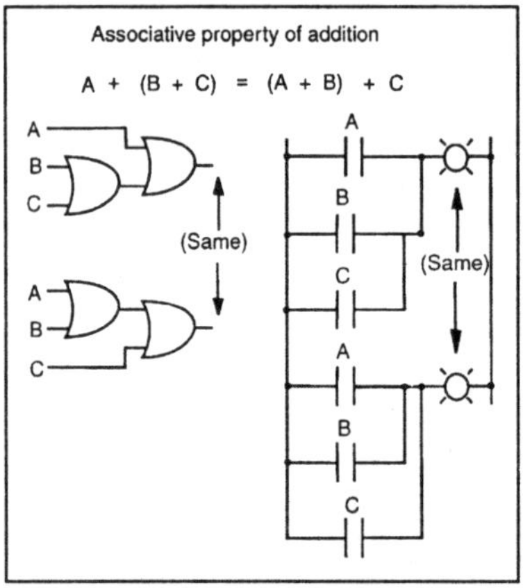


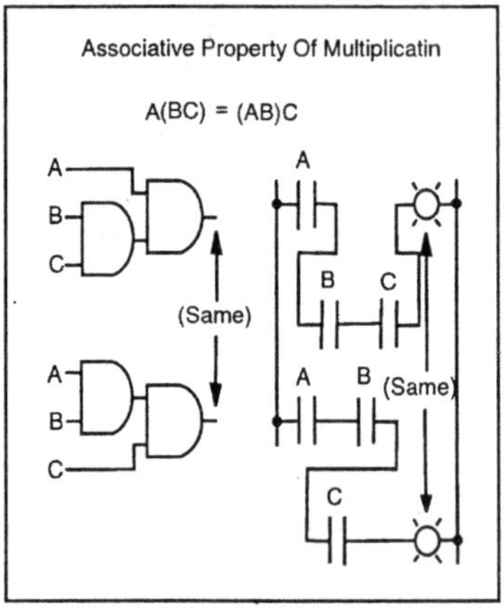


Lastly, we have the *distributive property*, illustrating how to expand a Boolean expression formed by the product of a sum, and in reverse shows us how terms may be factored out of Boolean sums-of-products:

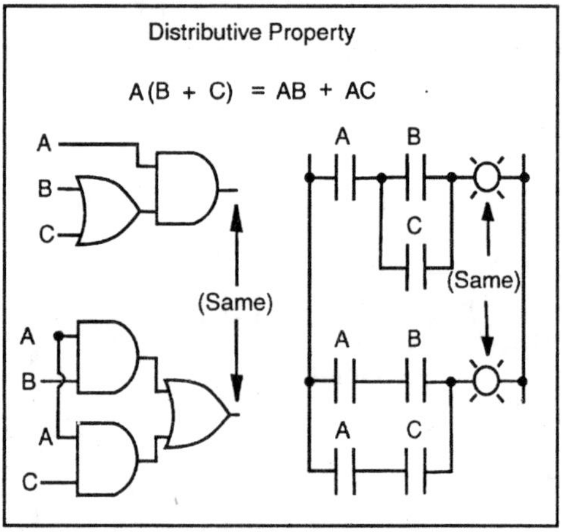


To summarize, here are the three basic properties: commutative, associative, and distributive.

**Table. Basic Bookean Algebraic Properties**

| Additive | Multiplicative |
|---|---|
| A + B = B + A | AB = BA |
| A + (B + C) = (A + B) + C | A(BC) = (AB) C |
| A (B + C) = AB + AC | |

## SIMPLIFICATION RULES

Boolean algebra finds its most practical use in the simplification of logic circuits. If we translate a logic circuit's function into symbolic (Boolean) form, and apply certain algebraic rules to the resulting equation to reduce the number of terms and/or arithmetic operations, the simplified equation may be translated back into circuit form for a logic circuit performing the same function with fewer components. If equivalent function may be achieved with fewer components, the result will be increased reliability and decreased cost of manufacture.To this end, there are several rules of Boolean algebra presented in this section for use in reducing expressions to their simplest forms.

The identities and properties already reviewed in this chapter are very useful in Boolean simplification, and for the most part bear similarity to many identities and properties of "normal" algebra. However, the rules shown in this section are all unique to Boolean mathematics.

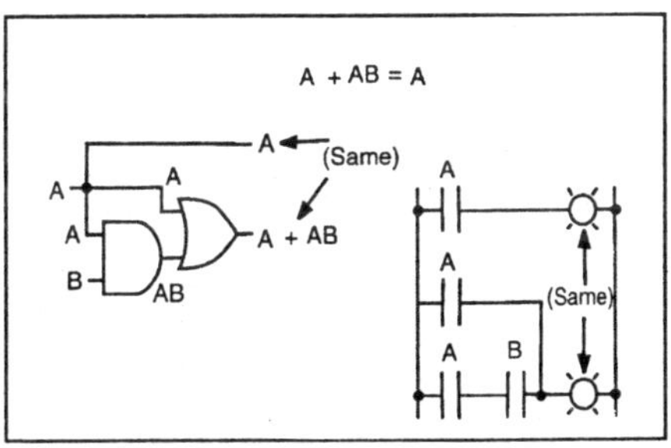


This rule may be proven symbolically by factoring an "A" out of the two terms, then applying the rules of A + 1 = 1 and 1A = A to achieve the final result:

$A + AB$

↓ Factoring A Out of Both Terms

$A(1 + B)$

↓ Applying identity $A + 1 = 1$

$A(1)$

↓ Applying identity $1A = A$

$A$

Please note how the rule A + 1 = 1 was used to reduce the (B + 1) term to 1. When a rule like "A + 1 = 1" is expressed using the letter "A", it doesn't mean it only applies to expressions containing "A". What the "A" stands for in a rule like A + 1 = 1 is *any* Boolean variable or collection of variables. This is perhaps the most difficult concept for new students to master in Boolean simplification: applying standardized identities, properties, and rules to expressions not in standard form. For instance, the Boolean expression ABC + 1 also reduces to 1 by means of the "A + 1 = 1" identity. In this case, we recognize that the "A" term in the identity's standard form can represent the entire "ABC" term in the original expression.

The next rule looks similar to the first one shown in this section, but is actually quite different and requires a more clever proof:

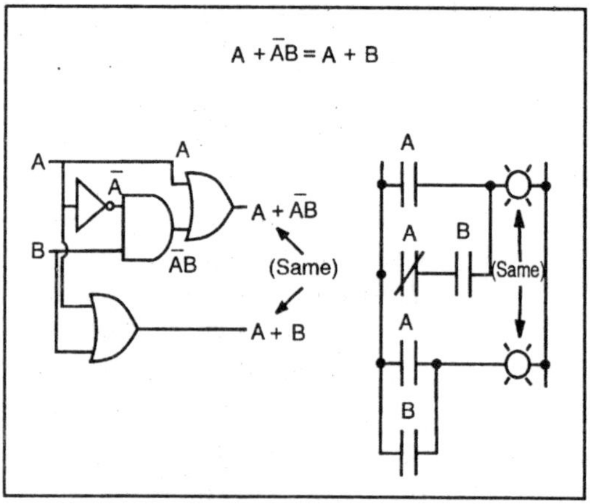


$A + \bar{A}B$

↓ Applying the Previous Rule to Expand A Term
A + AB = A

$A + AB + \bar{A}B$

↓ Factoring B Out of 2nd and 3rd Terms

$A + B(A + \bar{A})$

↓ Applying Identity A + A = 1

A + B (1)

↓ Applying Identity 1A = A

A + B

Note how the last rule (A + AB = A) is used to "un-simplify" the first "A" term in the expression, changing the "A" into an "A + AB". While this may seem like a backward step, it certainly helped to reduce the expression to something simpler.

Sometimes in mathematics we must take "backward" steps to achieve the most elegant solution. Knowing when to take such a step and when not to is part of the art-form of algebra, just as a victory in a game of chess almost always requires calculated sacrifices.

Another rule involves the simplification of a product-of-sums expression:

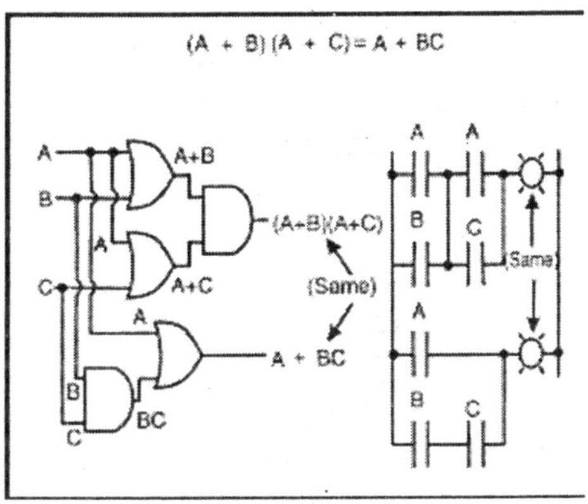


*And, the corresponding proof:*

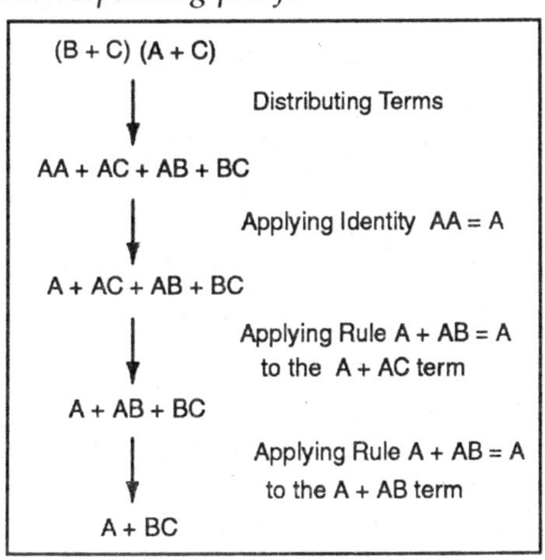

### *Useful Boolean Rules*

$$A + AB = A$$

$$A + \overline{A}B = A + B$$

$$(A + B)(A + C) = A + BC$$

## CIRCUIT SIMPLIFICATION

Let's begin with a semiconductor gate circuit in need of simplification. The "A," "B," and "C" input signals are assumed to be provided from switches, sensors, or perhaps other gate circuits. Where these signals originate is of no concern in the task of gate reduction.

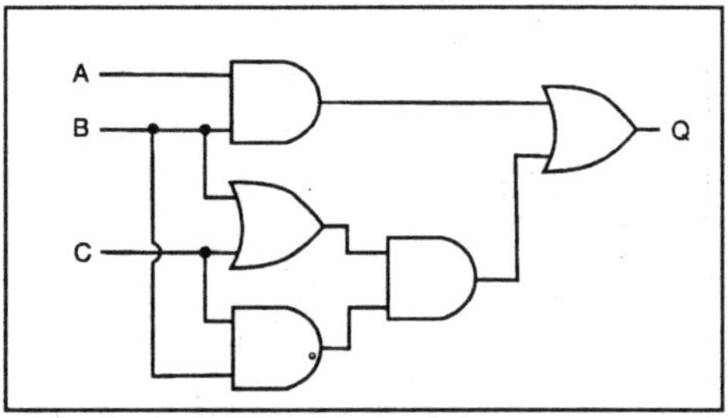


Our first step in simplification must be to write a Boolean expression for this circuit. This task is easily performed step by step if we start by writing sub-expressions at the output of each gate, corresponding to the respective input signals for each gate. Remember that OR gates are equivalent to Boolean addition, while AND gates are equivalent to Boolean multiplication. For example, I'll write sub-expressions at the outputs of the first three gates:

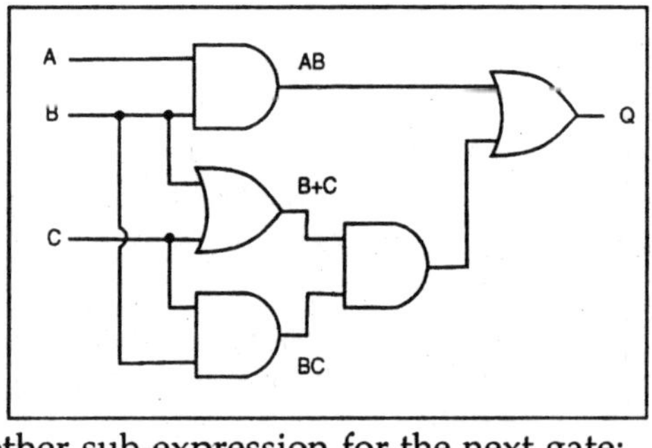


... then another sub-expression for the next gate:

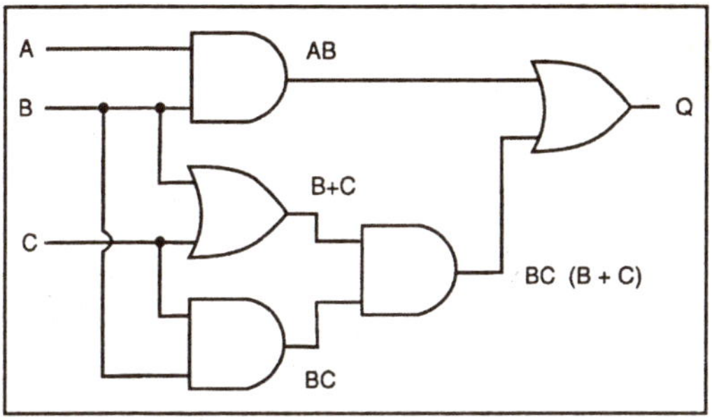


Finally, the output ("Q") is seen to be equal to the expression AB + BC(B + C):

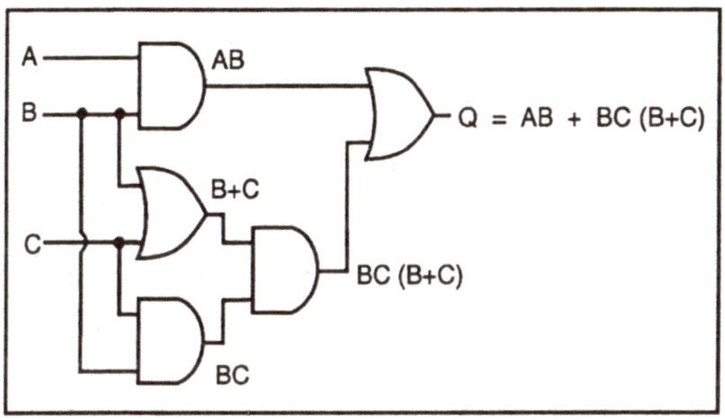


Now that we have a Boolean expression to work with, we need to apply the rules of Boolean algebra to reduce the expression to its simplest form (simplest defined as requiring the fewest gates to implement):

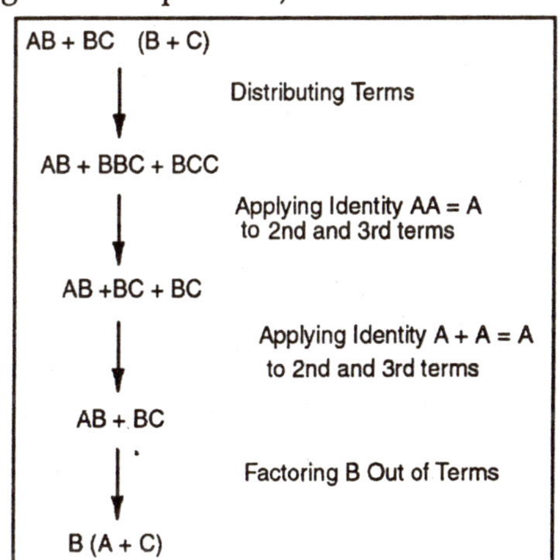

The final expression, B(A + C), is much simpler than the original, yet performs the same function. If you would like to verify this, you may generate a truth table for both expressions and determine Q's status (the circuits' output) for all eight logic-state combinations of A, B, and C, for both circuits. The two truth tables should be identical.

Now, we must generate a schematic diagram from this Boolean expression. To do this, evaluate the expression, following proper mathematical order of operations (multiplication before addition, operations inside parentheses before anything else), and draw gates for each step.

Remember again that OR gates are equivalent to Boolean addition, while AND gates are equivalent to Boolean multiplication. In this case, we would begin with the sub-expression "A + C", which is an OR gate:

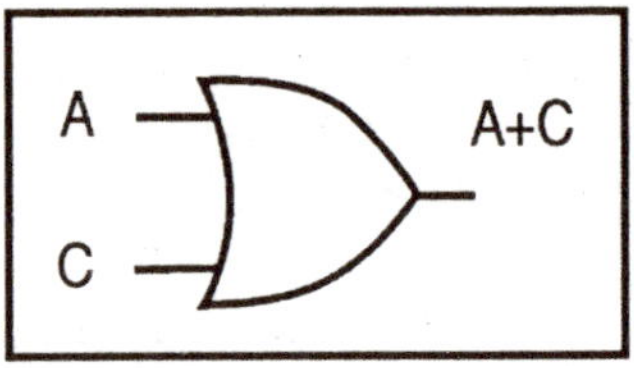


The next step in evaluating the expression "B(A + C)" is to multiply (AND gate) the signal B by the output of the previous gate (A + C):

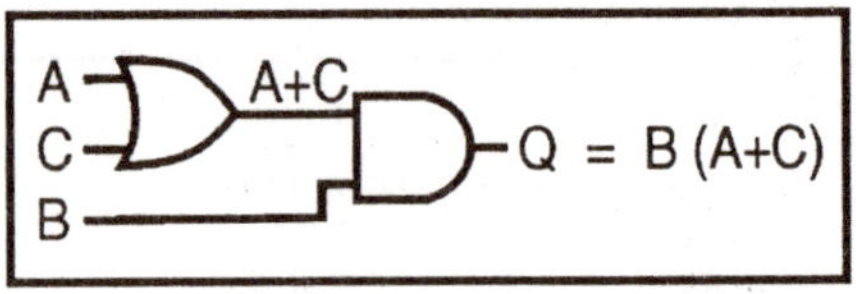


Obviously, this circuit is much simpler than the original, having only two logic gates instead of five. Such component reduction results in higher operating speed (less delay time from input signal transition to output signal transition), less power consumption, less cost, and greater reliability.

Electromechanical relay circuits, typically being slower, consuming more electrical power to operate, costing more, and having a shorter average life than their semiconductor counterparts, benefit dramatically from Boolean simplification. Let's consider an example circuit:

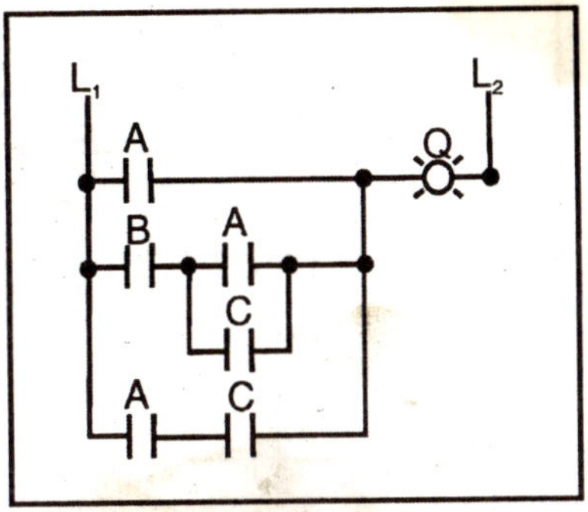


As before, our first step in reducing this circuit to its simplest form must be to develop a Boolean expression from the schematic. For example, examine the following resistor network with its resistors arranged in the same connection pattern as the relay contacts in the former circuit, and corresponding total resistance formula:

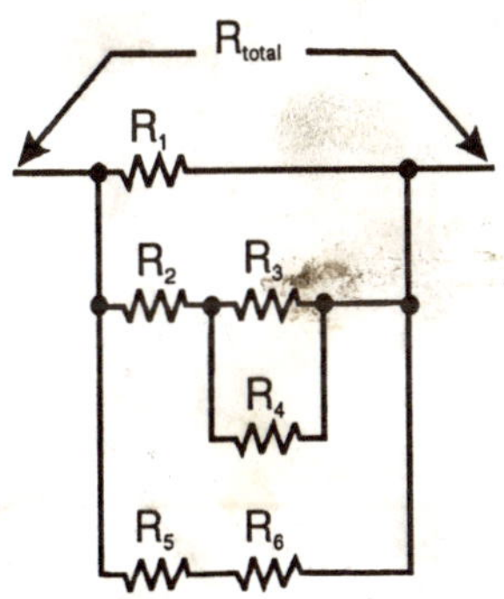


$$R_{total} = R_1 // [(R_3//R_4) -- R_2] // (R_5 -- R_6)$$

Remember that parallel contacts are equivalent to Boolean addition, while series contacts are equivalent to Boolean multiplication. Write a Boolean expression for this relay contact circuit, following the same order of precedence that you would follow in reducing a series-parallel resistor network to a total resistance. It may be helpful to write a Boolean sub-expression to the left of each ladder "rung," to help organize your expression-writing:

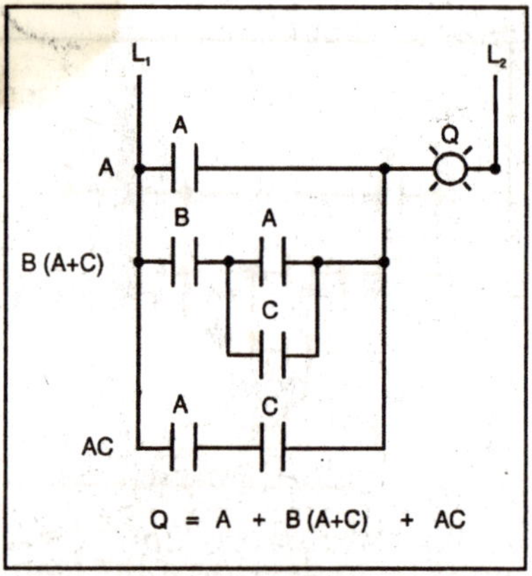


Now that we have a Boolean expression to work with, we need to apply the rules of Boolean algebra to reduce the expression to its simplest form (simplest defined as requiring the fewest relay contacts to implement):

A + B ( A + C) + AC
↓ Distributing Terms
A + AB + BC + AC
↓ Applying Rule A + AB = A to 1st and 2nd terms
A + BC + AC
↓ Applying Rule A + AB = A to 1st and 3rd terms
A + BC

The more mathematically inclined should be able to see that the two steps employing the rule "A + AB = A" may be combined into a single step, the rule being expandable to:

"A + AB + AC + AD +... = A"

A + B (A + C) + AC
↓ Distributing Terms
A + AB + BC + AC
↓ Applying (Expanded) Rule A + AB = A to 1st, 2nd, and 4th terms
A + BC

As you can see, the reduced circuit is much simpler than the original, yet performs the same logical function:

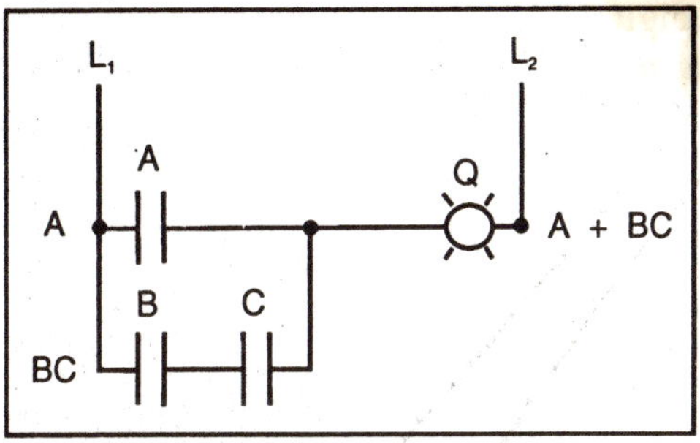


*Review:* To convert a gate circuit to a Boolean expression, label each gate output with a Boolean sub-expression corresponding to the gates' input signals, until a final expression is reached at the last gate.

To convert a Boolean expression to a gate circuit, evaluate the expression using standard order of operations: multiplication before addition, and operations within parentheses before anything else.

To convert a ladder logic circuit to a Boolean expression, label each rung with a Boolean sub-expression corresponding to the contacts' input signals, until a final expression is reached at the last coil or light. To determine proper order of evaluation, treat the contacts as though they were resistors, and as if you were determining total resistance of the series-parallel network formed by them.

In other words, look for contacts that are either *directly* in series or *directly* in parallel with each other first, then "collapse" them into equivalent Boolean sub-expressions before proceeding to other contacts.

To convert a Boolean expression to a ladder logic circuit, evaluate the expression using standard order of operations: multiplication before addition, and operations within parentheses before anything else.

## LOGIC CIRCUITS

Any logic circuit, no matter how complex, may be completely described using the Boolean operations, because the OR gate, AND gate, and NOT circuit are the basic building blocks of digital systems.

*This is an example of the circuit using Boolean expression:*

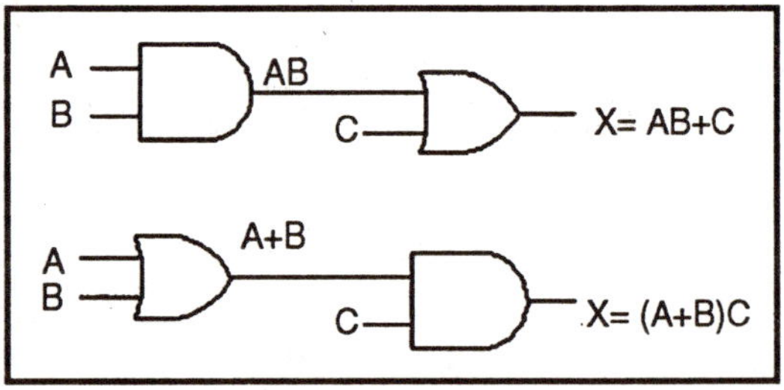


If an expression contains both AND and OR operations, the AND operations are performed first (X=AB+C: AB is performed first), unless there are parentheses in the expression, in which case the operation inside the parentheses is to be performed first (X=(A+B)+C: A+B is performed first).

## CIRCUITS CONTAINING INVERTERS

Whenever an INVERTER is present in a logic-circuit diagram, its output expression is simply equal to the input expression with a prime (′) over it.

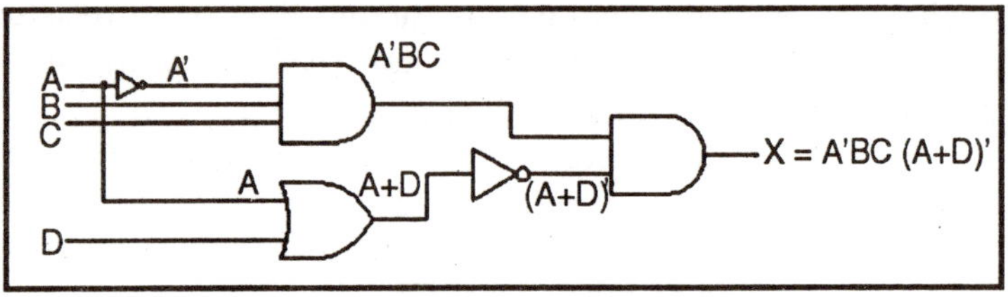


## EVALUATING LOGIC CIRCUIT OUTPUTS

Once the Boolean expression for a circuit output has been obtained, the output logic level can be determined for any set of input levels.

*This are two examples of the evaluating logic circuit output:*

Let A=0, B=1, C=1, D=1

X = A′BC (A+D)′

= 0′*1*1* (0+1)′

= 1 *1*1* (1)′
= 1 *1*1* 0
= 0

Let A=0, B=0, C=1, D=1, E=1

X = [D+ ((A+B)C)′] * E
= [1 + ((0+0)1 )′] * 1
= [1 + (0*1)′] * 1
= [1+ 0′] *1
= [1+ 1] * 1
= 1

In general, the following rules must always be followed when evaluating a Boolean expression:

- First, perform all inversions of single terms; that is, 0 = 1 or 1 = 0.
- Then perform all operations within parentheses.
- Perform an AND operation before an OR operation unless parentheses indicate otherwise.
- If an expression has a bar over it, perform the operations of the expression first and then invert the result.

## DETERMINING OUTPUT LEVEL

The output logic level for given input levels can also be determined directly from the circuit diagram without using the Boolean expression.

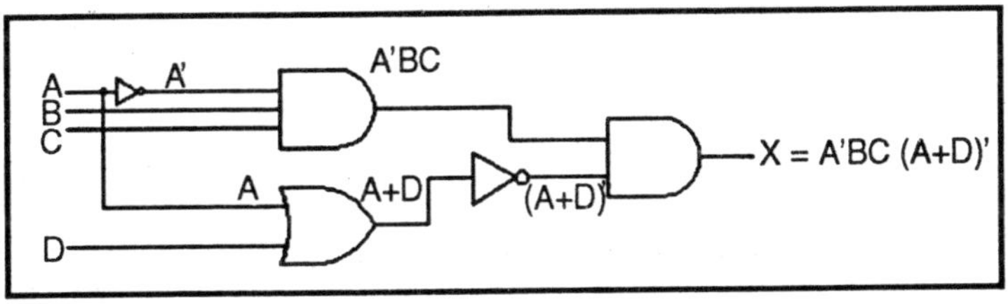


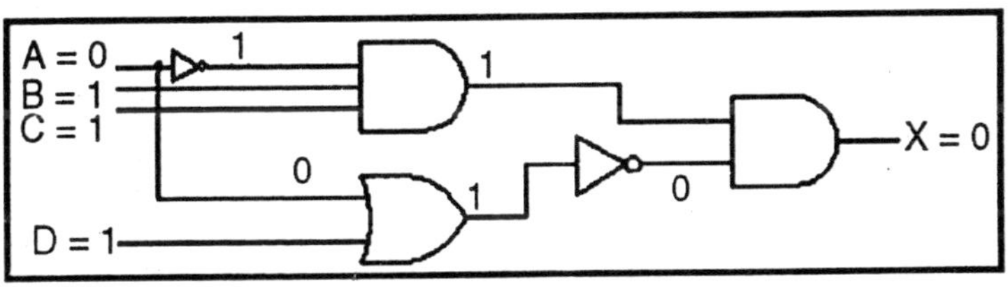

## IMPLEMENTING CIRCUITS FROM BOOLEAN EXPRESSION

If the operation of a circuit is defined by a Boolean expression, a logic-circuit diagram can he implemented directly from that expression. Suppose that we wanted to construct a circuit whose output is y = AC+BC′ + A′BC. This Boolean expression contains three terms (AC, BC′, A′BC), which are ORed together. This tells us that a three-input OR gate is required with inputs that are equal to AC, BC′, and A′BC, respectively.

Each OR-gate input is an AND product term, which means that an AND gate with appropriate inputs can be used to generate each of these terms. Note the use of INVERTERs to produce the A′ and C′ terms required in the expression.

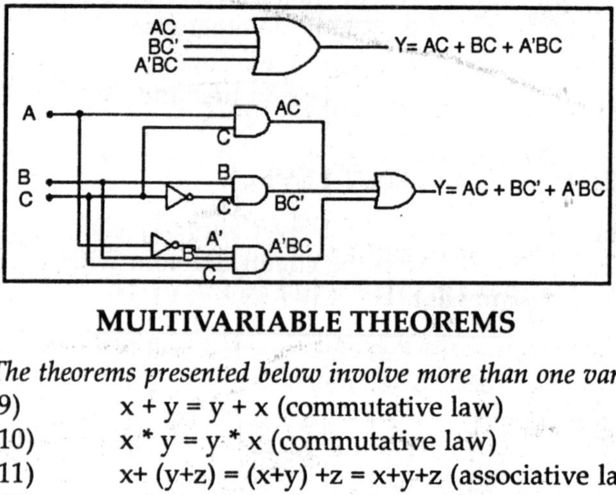


## MULTIVARIABLE THEOREMS

*The theorems presented below involve more than one variable:*

(9) x + y = y + x (commutative law)

(10) x * y = y * x (commutative law)

(11) x+ (y+z) = (x+y) +z = x+y+z (associative law)

(12) x (yz) = (xy) z = xyz (associative law)

(13a) x (y+z) = xy + xz

(13b) (w+x)(y+z) = wy + xy + wz + xz

(14) x + xy = x [proof see below]

(15) x + x′y = x + y

Proof of (14)

x + xy = x (1+y)

= x * 1 [using theorem (6)]

= x [using theorem (2)]

## BOOLEAN THEOREMS

Investigating the various Boolean theorems (rules) can help us to simplify logic expressions and logic circuits.

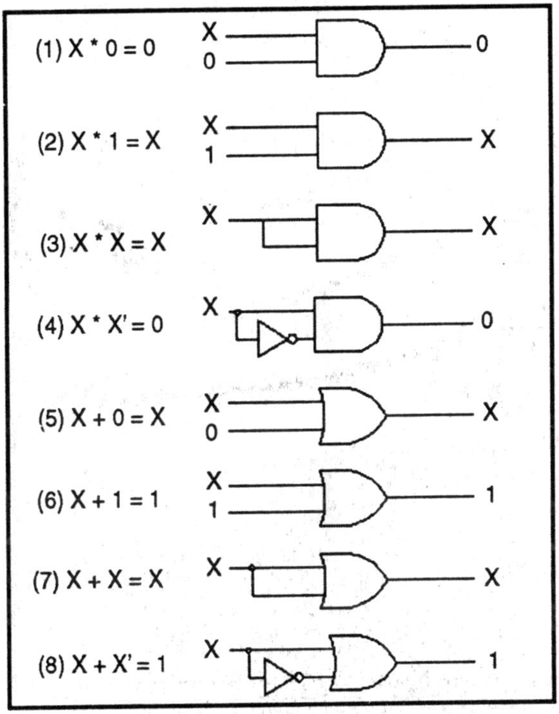


## DEMORGAN'S THEOREM

DeMorgan's theorems are extremely useful in simplifying expressions in which a product or sum of variables is inverted. The two theorems are:

(16) $(x+y)' = x' * y'$

(17) $(x*y)' = x' + y'$

Theorem (16) says that when the OR sum of two variables is inverted, this is the same as inverting each variable individually and then ANDing these inverted variables.

Theorem (17) says that when the AND product of two variables is inverted, this is the same as inverting each variable individually and then ORing them.

*Example*:

X = [(A′+C) * (B+D′)]′
= (A′+C)′ + (B+D′)′ [by theorem (17)]
= (A″*C′) + (B′+D″) [by theorem (16)]
= AC′ + B′D

*Three Variables DeMorgan's Theorem*:

(18) (x+y+z)′ = x′ * y′ * z′

(19) (xyz)′ = x′ + y′ + z′

*Implications of DeMorgan's Theorem*:

For (16): (x+y)′ = x′ * y′

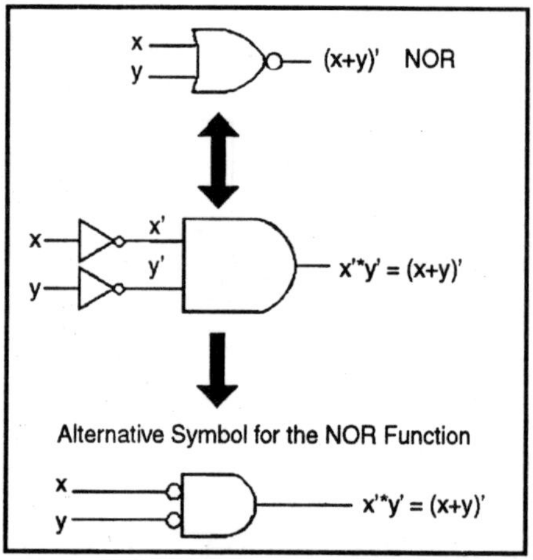


For (17): (x*y)′ = x′ + y′

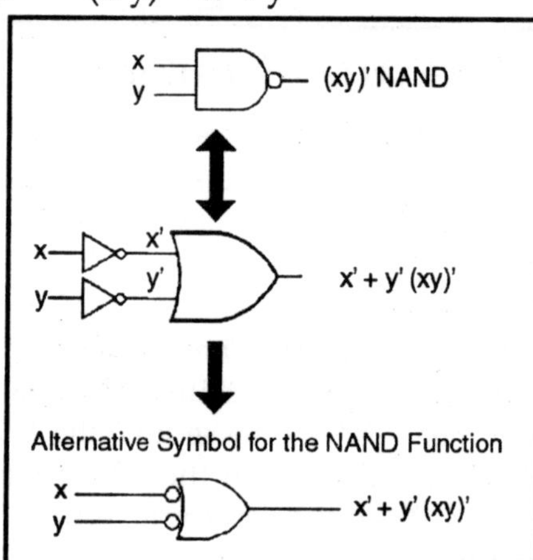

## UNIVERSALITY OF NAND & NOR GATES

It is possible to implement any logic expression using only NAND gates and no other type of gate. This is because NAND gates, in the proper combination, can be used to perform each of the Boolean operations OR, AND, and INVERT.

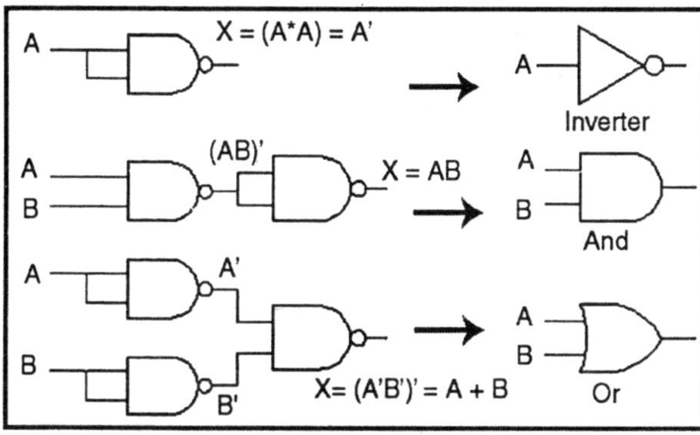


In a similar manner, it can be shown that NOR gates can be arranged to implement any of the Boolean operations.

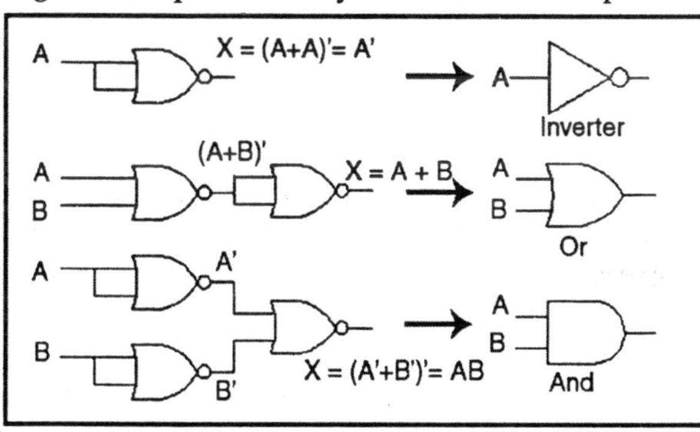


## LOGIC GATE REPRESENTATIONS

The left side of the illustration shows the standard symbol for each logic gate, and the right side shows the alternate symbol. The alternate symbol for each gate is obtained from the standard symbol by doing the following:

- Invert each input and output of the standard symbol. This is done by adding bubbles (small circles) on input and output lines that do not have bubbles, and by removing bubbles that are already there.

- Change the operation symbol from AND to OR, or from OR to AND. (In the special case of the INVERTER, the operation symbol is not changed.)

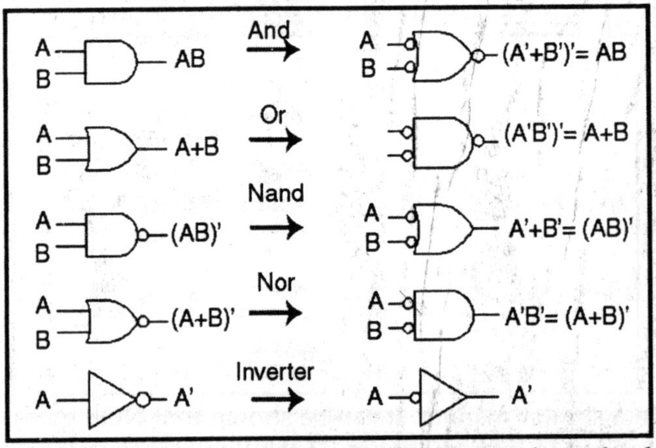


Several points should be stressed regarding the logic symbol equivalences:

- The equivalences are valid for gates with any number of inputs.
- None of the standard symbols have bubbles on their inputs, and all the alternate symbols do.
- The standard and alternate symbols for each gate represent the same physical circuit: there is no difference in the circuits represented by the two symbols.
- NAND and NOR gates are inverting gates, and so both the standard and alternate symbols for each will have a bubble on either the input or the output. AND and OR gates are noninverting gates, and so the alternate symbols for each will have bubbles on both inputs and output.

## LOGIC SYMBOL INTERPRETATION

### Active Logic Levels

When an input or output line on a logic circuit symbol

has no bubble on it, that line is said to be active-HIGH. When an input or output line does have a bubble on it, that line is said to be active-LOW. The presence or absence of a bubble, then, determines the active-HIGH/active-LOW status of a circuit's inputs and output, and is used to interpret the circuit operation.

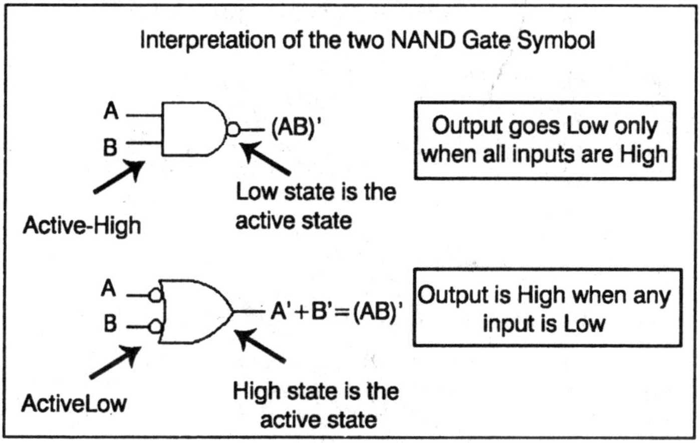


## BINARY SYSTEM

### CONCEPTS

To understand binary numbers, begin by recalling elementary school math. When we first learned about numbers, we were taught that, in the decimal system, things are organized into columns:

$$\begin{array}{c|c|c} H & T & O \\ 1 & 9 & 3 \end{array}$$

Such that "H" is the hundreds column, "T" is the tens column, and "O" is the ones column. So the number "193" is 1-hundreds plus 9-tens plus 3-ones. later, we learned that the ones column meant 10^0, the tens column meant 10^1, the hundreds column 10^2 and so on, such that

$$\begin{array}{c|c|c} 10^2 & 10^1 & 10^0 \\ 1 & 9 & 3 \end{array}$$

The number 193 is really {(1*10^2)+(9*10^1)+(3*10^0)}.

As you know, the decimal system uses the digits 0-9 to represent numbers. If we wanted to put a larger number in column 10^n (e.g., 10), we would have to multiply 10*10^n, which would give 10^(n+1), and be carried a column to the left. For example, putting ten in the 10^0 column is impossible, so we put a 1 in the 10^1 column, and a 0 in the 10^0 column, thus using two columns. Twelve would be 12*10^0, or 10^0(10+2), or 10^1+2*10^0, which also uses an additional column to the left (12). The binary system works under the exact same principles as the decimal system, only it operates in base 2 rather than base 10. In other words, instead of columns being

10 ^ 2 | 10 ^ 1 | 10 ^ 0 they are

2 ^ 2 | 2 ^ 1 | 2 ^ 0

Instead of using the digits 0-9, we only use 0-1 (again, if we used anything larger it would be like multiplying 2*2^n and getting 2^n+1, which would not fit in the 2^n column. Therefore, it would shift you one column to the left. For example, "3" in binary cannot be put into one column.

The first column we fill is the right-most column, which is 2^0, or 1. Since 3>1, we need to use an extra column to the left, and indicate it as "11" in binary (1*2^1) + (1*2^0).

## BINARY ADDITION

Adding binary numbers is a very simple task, and very similar to the longhand addition of decimal numbers. As with decimal numbers, you start by adding the bits (digits) one column, or place weight, at a time, from right to left. Unlike decimal addition, there is little to memorize in the way of rules for the addition of binary bits:

$$0 + 0 = 0$$
$$1 + 0 = 1$$
$$0 + 1 = 1$$
$$1 + 1 = 10$$
$$1 + 1 + 1 = 11$$

Just as with decimal addition, when the sum in one column is a two-bit (two-digit) number, the least significant figure is written as part of the total sum and the most significant figure is "carried" to the next left column. Consider the following examples:

$$\begin{array}{ccc} & 11\ 1 \leftarrow \text{Carry bits} \rightarrow 11 & \\ 1001101 & 1001001 & 1000111 \\ +0010010 & +0011001 & +0010110 \\ \hline 1011111 & 1100010 & 1011101 \end{array}$$

The addition problem on the left did not require any bits to be carried, since the sum of bits in each column was either 1 or 0, not 10 or 11. In the other two problems, there definitely were bits to be carried, but the process of addition is still quite simple. As we'll see later, there are ways that electronic circuits can be built to perform this very task of addition, by representing each bit of each binary number as a voltage signal (either "high," for a 1; or "low" for a 0). This is the very foundation of all the arithmetic which modern digital computers perform.

*Consider the addition of decimal numbers:*

$$\begin{array}{r} 23 \\ +48 \\ \hline \end{array}$$

We begin by adding 3+8=11. Since 11 is greater than 10, a one is put into the 10's column (carried), and a 1 is recorded in the one's column of the sum. Next, add {(2+4) +1} (the one is from the carry)=7, which is put in the 10's column of the sum. Thus, the answer is 71.

Binary addition works on the same principle, but the numerals are different. Begin with one-bit binary addition:

$$\begin{array}{ccc} 0 & 0 & 1 \\ +0 & +1 & +0 \\ \hline 0 & 1 & 1 \end{array}$$

1+1 carries us into the next column. In decimal form, 1+1=2. In binary, any digit higher than 1 puts us a column to the left (as would 10 in decimal notation). The decimal number "2" is written in binary notation as "10" (1*2^1)+(0*2^0). Record the 0 in the ones column, and carry the 1 to the twos column to get an answer of "10." In our vertical notation,

$$\begin{array}{r} 1 \\ +1 \\ \hline 10 \end{array}$$

*The process is the same for multiple-bit binary numbers:*

$$\begin{array}{r} 1010 \\ +1111 \\ \hline \end{array}$$

- *Step one*: Column 2^0: 0+1=1.
  Record the 1.
  *Temporary Result*: 1; Carry: 0
- *Step two*: Column 2^1: 1+1=10.
  Record the 0, carry the 1.
  *Temporary Result*: 01; Carry: 1
- *Step three*: Column 2^2: 1+0=1 Add 1 from carry: 1+1=10.
  Record the 0, carry the 1.
- *Step four*: Column 2^3: 1+1=10. Add 1 from carry: 10+1=11.
  Record the 11.
  *Final result*: 11001

**Alternately**

$$\begin{array}{r} 11(\text{carry}) \\ 1010 \\ +1111 \\ \hline 11001 \end{array}$$

*Always remember*:

- 0+0=0
- 1+0=1
- 1+1=10

*Try a few examples of binary addition:*

$$\begin{array}{rrr} 111 & 101 & 111 \\ +110 & +111 & +111 \\ \hline \end{array}$$

## Numbers Versus Numeration

It is imperative to understand that the type of numeration system used to represent numbers has no impact upon the outcome of any arithmetical function (addition, subtraction, multiplication, division, roots, powers, or logarithms). A number is a number is a number; one plus one will always equal two (so long as we're dealing with real numbers), no matter how you symbolize one, one, and two. A prime number in decimal form is still prime if its shown in binary form, or octal, or hexadecimal. Π is still the ratio between the circumference and diameter of a circle, no matter what symbols you use to denote its value.

The essential functions and interrelations of mathematics are unaffected by the particular system of symbols we might choose to represent quantities. This distinction between numbers and systems of numeration is critical to understand. The essential distinction between the two is much like that between an object and the spoken word(s) we associate with it. A house is still a house regardless of whether we call it by its English name house or its Spanish name casa. The first is the actual thing, while the second is merely the symbol for the thing.

That being said, performing a simple arithmetic operation such as addition (longhand) in binary form can be confusing to a person accustomed to working with decimal numeration only. In this lesson, we'll explore the techniques used to perform simple arithmetic functions on binary numbers, since these techniques will be employed in the design of electronic circuits to do the same.

You might take longhand addition and subtraction for granted, having used a calculator for so long, but deep inside that calculator's circuitry all those operations are performed "longhand," using binary numeration. To understand how that's accomplished, we need to review to the basics of arithmetic.

## Negative Binary Numbers

With addition being easily accomplished, we can perform the operation of subtraction with the same technique simply by making one of the numbers negative. For example, the subtraction problem of 7 - 5 is essentially the same as the addition problem 7 + (-5). Since we already know how to represent positive numbers in binary, all we need to know now is how to represent their negative counterparts and we'll be able to subtract. Usually we represent a negative decimal number by placing a minus sign directly to the left of the most significant digit, just as in the example above, with -5. However, the whole purpose of using binary notation is for constructing on/off circuits that can represent bit values in terms of voltage (2 alternative values: either "high" or "low"). In this context, we don't have the luxury of a third symbol such as a "minus" sign, since these circuits can only be on or off (two possible states). One solution is to reserve a bit (circuit) that does nothing but represent the mathematical sign:

$$101_2 = 5_{10} \text{ (positive)}$$

*Extra bit, representing sign (0=positive, 1=negative):*

|

$$0101_2 = 5_{10} \text{ (positive)}$$

*Extra bit, representing sign (0=positive, 1=negative):*

|

$$1101_2 = -5_{10} \text{ (negative)}$$

As you can see, we have to be careful when we start using bits for any purpose other than standard place-weighted values. Otherwise, $1101_2$ could be misinterpreted as the number thirteen when in fact we mean to represent negative five. To keep things straight here, we must first decide how many bits are going to be needed to represent the largest numbers we'll be dealing with, and then be sure not to exceed that bit field length in our arithmetic operations.

For the above example, I've limited myself to the representation of numbers from negative seven ($1111_2$) to positive seven ($0111_2$), and no more, by making the fourth bit the "sign" bit. Only by first establishing these limits can I avoid confusion of a negative number with a larger, positive number.

Representing negative five as $1101_2$ is an example of the sign-magnitude system of negative binary numeration. By using the leftmost bit as a sign indicator and not a place-weighted value, I am sacrificing the "pure" form of binary notation for something that gives me a practical advantage:

the representation of negative numbers. The leftmost bit is read as the sign, either positive or negative, and the remaining bits are interpreted according to the standard binary notation: left to right, place weights in multiples of two. As simple as the sign-magnitude approach is, it is not very practical for arithmetic purposes.

There's another method for representing negative numbers which works with our familiar technique of longhand addition, and also happens to make more sense from a place-weighted numeration point of view, called complementation. With this strategy, we assign the leftmost bit to serve a special purpose, just as we did with the sign-magnitude approach, defining our number limits just as before. However, this time, the leftmost bit is more than just a sign bit; rather, it possesses a negative place-weight value. For example, a value of negative five would be represented as such:

Extra bit, place weight = Negative eight

|

$$1011_2 = 5_{10}\,(\text{negative})$$

$$(1 \times -8_{10}) + (0 \times 4_{10}) + (1 \times 2_{10}) + (1 \times 1_{10}) = -5_{10}$$

With the right three bits being able to represent a magnitude from zero through seven, and the leftmost bit representing either zero or negative eight, we can successfully represent any integer number from negative seven ($1001_2 = -8_{10} + 1_{10} = -7_{10}$) to positive seven ($0111_2 = 0_{10} + 7_{10} = 7_{10}$).

Representing positive numbers in this scheme (with the fourth bit designated as the negative weight) is no different from that of ordinary binary notation. However, represent-ing negative numbers is not quite as straightforward:

| | | | |
|---|---|---|---|
| Zero | 0000 | Negative eight | 1000 |
| Positive one | 0001 | Negative one | 1111 |
| Positive two | 0010 | Negative two | 1110 |
| Positive three | 0011 | Negative three | 1101 |
| Positive four | 0100 | Negative four | 1100 |
| Positive five | 0101 | Negative five | 1011 |
| Positive six | 0110 | Negative six | 1010 |
| Positive seven | 0111 | Negative seven | 1001 |

Note that the negative binary numbers in the right column, being the sum of the right three bits' total plus the negative eight of the leftmost bit, don't "count" in the same progression as the positive binary numbers in the left column. Rather, the right three bits have to be set at the proper value to equal the desired (negative) total when summed with the negative eight place value of the leftmost bit.

Those right three bits are referred to as the two's complement of the corresponding positive number. Consider the following comparison:

| Positive Number | Two's Complement |
|---|---|
| 001 | 111 |
| 010 | 110 |
| 011 | 101 |
| 100 | 100 |
| 101 | 011 |
| 110 | 010 |
| 111 | 001 |

In this case, with the negative weight bit being the fourth bit (place value of negative eight), the two's complement for any positive number will be whatever value is needed to add to negative eight to make that positive value's negative equivalent.

Thankfully, there's an easy way to figure out the two's complement for any binary number: simply invert all the bits of that number, changing all 1's to 0's and vice versa (to arrive at what is called the one's complement) and then add one. For example, to obtain the two's complement of five ($101_2$), we would first invert all the bits to obtain $010_2$ (the "one's complement"), then add one to obtain $011_2$, or $-5_{10}$ in three-bit, two's complement form.

Interestingly enough, generating the two's complement of a binary number works the same if you manipulate all the bits, including the leftmost (sign) bit at the same time as the magnitude bits. Let's try this with the former example, converting a positive five to a negative five, but performing the complementation process on all four bits.

We must be sure to include the 0 (positive) sign bit on the original number, five ($0101_2$). First, inverting all bits to obtain the one's complement: $1010_2$. Then, adding one, we obtain the final answer: $1011_2$, or $-5_{10}$ expressed in four-bit, two's complement form.

It is critically important to remember that the place of the negative-weight bit must be already determined before any two's complement conversions can be done. If our binary numeration field were such that the eighth bit was designated as the negative-weight bit ($10000000_2$), we'd have to determine the two's complement based on all seven of the other bits.

Here, the two's complement of five ($0000101_2$) would be $1111011_2$. A positive five in this system would be represented as $00000101_2$, and a negative five as $11111011_2$.

## BINARY SUBTRACTION

We can subtract one binary number from another by using the standard techniques adapted for decimal numbers (subtraction of each bit pair, right to left, "borrowing" as needed from bits to the left). However, if we can leverage the already familiar (and easier) technique of binary addition to subtract, that would be better.

As we just learned, we can represent negative binary numbers by using the "two's complement" method and a negative place-weight bit. Here, we'll use those negative binary numbers to subtract through addition. Here's a sample problem:

*Subtraction*: $7_{10} - 5_{10}$ Addition equivalent: $7_{10} + (-5_{10})$

If all we need to do is represent seven and negative five in binary (two's complemented) form, all we need is three bits plus the negative-weight bit:

Positive seven = $0111_2$

Negative five = $1011_2$

Now, let's add them together:

```
 1111  ←Carry bits
  0111
 +1011
 -----
 10010
```

|
Discard extra bit
Answer = $0010_2$

Since we've already defined our number bit field as three bits plus the negative-weight bit, the fifth bit in the answer (1) will be discarded to give us a result of $0010_2$, or positive two, which is the correct answer.

Another way to understand why we discard that extra bit is to remember that the leftmost bit of the lower number possesses a negative weight, in this case equal to negative eight. When we add these two binary numbers together, what we're actually doing with the MSBs is subtracting the lower number's MSB from the upper number's MSB. In subtraction, one never "carries" a digit or bit on to the next left place-weight.

Let's try another example, this time with larger numbers. If we want to add $-25_{10}$ to $18_{10}$, we must first decide how large our binary bit field must be. To represent the largest (absolute value) number in our problem, which is twenty-five, we need at least five bits, plus a sixth bit for the negative-weight bit. Let's start by representing positive twenty-five, then finding the two's complement and putting it all together into one numeration:

$+25_{10} = 011001_2$ (showing all six bits)
One's complement of $11001_2 = 100110_2$
One's complement + 1 = two's complement = $100111_2$
$-25_{10} = 100111_2$

Essentially, we're representing negative twenty-five by using the negative-weight (sixth) bit with a value of negative thirty-two, plus positive seven (binary $111_2$).

Now, let's represent positive eighteen in binary form, showing all six bits: $18_{10} = 010010_2$

*Now, let's add them together and see what we get:*

```
11  ← Carry bits
1001111
+010010
-------
111001
```

Since there were no "extra" bits on the left, there are no bits to discard. The leftmost bit on the answer is a 1, which means that the answer is negative, in two's complement form, as it should be. Converting the answer to decimal form by summing all the bits times their respective weight values, we get:

$(1 \times -32_{10}) + (1 \times 16_{10}) + (1 \times 8_{10}) + (1 \times 1_{10}) = -7_{10}$

Indeed $-7_{10}$ is the proper sum of $-25_{10}$ and $18_{10}$.

## Overflow

One caveat with signed binary numbers is that of overflow, where the answer to an addition or subtraction problem exceeds the magnitude which can be represented with the alloted number of bits. Remember that the place of the sign bit is fixed from the beginning of the problem.

With the last example problem, we used five binary bits to represent the magnitude of the number, and the left-most (sixth) bit as the negative-weight, or sign, bit. With five bits to represent magnitude, we have a representation range of $2^5$, or thirty-two integer steps from 0 to maximum.

This means that we can represent a number as high as $+31_{10}$ ($011111_2$), or as low as $-32_{10}$ ($100000_2$). If we set up an addition problem with two binary numbers, the sixth bit used for sign, and the result either exceeds $+31_{10}$ or is less than -$32_{10}$, our answer will be incorrect. Let's try adding $17_{10}$ and $19_{10}$ to see how this overflow condition works for excessive positive numbers:

$17_{10} = 10001_2$ $\qquad\qquad$ $19_{10} = 10011_2$

$$\begin{array}{ll} 1\,11 & \leftarrow \text{Carry bits} \\ \text{(Showing sign bits)} & 010001 \\ +1010011 & \\ \hline 100100 & \end{array}$$

The answer ($100100_2$), interpreted with the sixth bit as the $-32_{10}$ place, is actually equal to $-28_{10}$, not $+36_{10}$ as we should get with $+17_{10}$ and $+19_{10}$ added together! Obviously, this is not correct. What went wrong? The answer lies in the restrictions of the six-bit number field within which we're working Since the magnitude of the true and proper sum ($36_{10}$) exceeds the allowable limit for our designated bit field, we have an overflow error.

Simply put, six places doesn't give enough bits to represent the correct sum, so whatever figure we obtain using the strategy of discarding the left-most "carry" bit will be incorrect.

A similar error will occur if we add two negative numbers together to produce a sum that is too low for our six-bit binary field. Let's try adding -$17_{10}$ and -$19_{10}$ together to see how this works (or doesn't work, as the case may be!):

```
                1 1111    ← Carry bits
(Showing sign bits)  101111
                    +010011
                    -------
                    1011100
                    |
         Discard extra bit
```

Final Answer: $011100_2 = +28_{10}$

The (incorrect) answer is a positive twenty-eight. The fact that the real sum of negative seventeen and negative nineteen was too low to be properly represented with a five bit magnitude field and a sixth sign bit is the root cause of this difficulty.

Let's try these two problems again, except this time using the seventh bit for a sign bit, and allowing the use of 6 bits for representing the magnitude:

$17_{10} + 19_{10}$ $\qquad\qquad$ $(-17_{10}) + (-19_{10})$

```
   1 11        11 1111
  0010001     1101111
+ 0010011   + 1101101
---------   ---------
01001002    110111002
              |
```

ANSWERS: $0100100_2 = +36_{10}$

$1011100_2 = -36_{10}$

By using bit fields sufficiently large to handle the magnitude of the sums, we arrive at the correct answers. In these sample problems we've been able to detect overflow errors by performing the addition problems in decimal form and comparing the results with the binary answers. For example, when adding $+17_{10}$ and $+19_{10}$ together, we knew that the answer was supposed to be $+36_{10}$, so when the binary sum checked out to be $-28_{10}$, we knew that something had to be wrong. Although this is a valid way of detecting overflow, it is not very efficient. After all, the whole idea of complementation is to be able to reliably add binary numbers together and not have to double-check the result by adding the same numbers together in decimal form.

This is especially true for the purpose of building electronic circuits to add binary quantities together: the circuit has to be able to check itself for overflow without the supervision of a human being who already knows what the correct answer is. What we need is a simple error-detection method that doesn't require any additional arithmetic. Perhaps the most elegant solution is to check for the sign of the sum and compare it against the signs of the numbers added. Obviously, two positive numbers added together should give a positive result, and two negative numbers added together should give a negative result.

Notice that whenever we had a condition of overflow in the example problems, the sign of the sum was always opposite of the two added numbers: $+17_{10}$ plus $+19_{10}$ giving $-28_{10}$, or $-17_{10}$ plus $-19_{10}$ giving $+28_{10}$. By checking the signs alone we are able to tell that something is wrong.

But what about cases where a positive number is added to a negative number? What sign should the sum be in order to be correct. Or, more precisely, what sign of sum would necessarily indicate an overflow error? The answer to this is equally elegant: there will never be an overflow error when two numbers of opposite signs are added together! The reason for this is apparent when the nature of overflow is considered. Overflow occurs when the magnitude of a number exceeds the range allowed by the size of the bit field. The sum of two identically-signed numbers may very well exceed the range of the bit field of those two numbers, and so in this case overflow is a possibility. However, if a positive number is added to a negative number, the sum will always be closer to zero than either of the two added numbers: its magnitude must be less than the magnitude of either original number, and so overflow is impossible.

Fortunately, this technique of overflow detection is easily implemented in electronic circuitry, and it is a standard feature in digital adder circuits: a subject for a later chapter.

## Bit Groupings

The singular reason for learning and using the binary numeration system in electronics is to understand how to design, build, and troubleshoot circuits that represent and process numerical quantities in digital form. Since the bivalent (two-valued) system of binary bit numeration lends itself so easily to representation by "on" and "off" transistor states (saturation and cutoff, respectively), it makes sense to design and build circuits leveraging this principle to perform binary calculations. If we were to build a circuit to represent a binary number, we would have to allocate enough transistor circuits to represent as many bits as we desire. In other words, in designing a digital circuit, we must first decide how many bits (maximum) we would like to be able to represent, since each bit requires one on/off circuit to represent it. This is analogous to designing an abacus to digitally represent decimal numbers: we must decide how many digits we wish to handle in this primitive "calculator" device, for each digit requires a separate rod with its own beads.

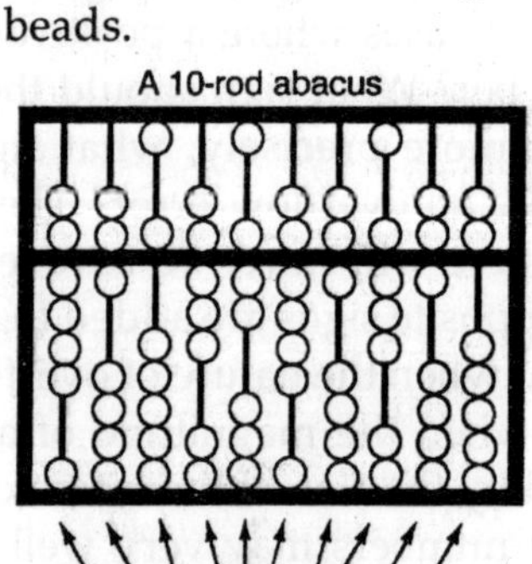

A ten-rod abacus would be able to represent a ten-digit decimal number, or a maximum value of 9,999,999,999. If we wished to represent a larger number on this abacus, we would be unable to, unless additional rods could be added to it.

In digital, electronic computer design, it is common to design the system for a common "bit width:" a maximum number of bits allocated to represent numerical quantities. Early digital computers handled bits in groups of four or eight. More modern systems handle numbers in clusters of 32 bits or more.

To more conveniently express the "bit width" of such clusters in a digital computer, specific labels were applied to the more common groupings. Eight bits, grouped together to form a single binary quantity, is known as a byte. Four bits, grouped together as one binary number, is known by the humorous title of nibble, often spelled as nybble.

A multitude of terms have followed byte and nibble for labeling specfiic groupings of binary bits. Most of the terms shown here are informal, and have not been made "authoritative" by any standards group or other sanctioning body. However, their inclusion into this chapter is warranted by their occasional appearance in technical literature, as well as the levity they add to an otherwise dry subject:

- *Bit:* A single, bivalent unit of binary notation. Equivalent to a decimal "digit."
- *Crumb, Tydbit, or Tayste:* Two bits.
- *Nibble, or Nybble:* Four bits.
- *Nickle:* Five bits.
- *Byte:* Eight bits.
- *Deckle:* Ten bits.
- *Playte:* Sixteen bits.
- *Dynner:* Thirty-two bits.
- *Word:* (system dependent).

The most ambiguous term by far is word, referring to the standard bit-grouping within a particular digital system. For a computer system using a 32 bit-wide "data path," a "word" would mean 32 bits. If the system used 16 bits as the standard grouping for binary quantities, a "word" would mean 16 bits.

The terms playte and dynner, by contrast, always refer to 16 and 32 bits, respectively, regardless of the system context in which they are used. Context dependence is likewise true for derivative terms of word, such as double word and longword (both meaning twice the standard bit-width), half-word (half the standard bit-width), and quad (meaning four times the standard bit-width). One humorous addition to this somewhat boring collection of word-derivatives is the term chawmp, which means the same as half-word. For example, a chawmp would be 16 bits in the context of a 32-bit digital system, and 18 bits in the context of a 36-bit system. Also, the term gawble is sometimes synonymous with word.

## BINARY MULTIPLICATION

Multiplication in the binary system works the same way as in the decimal system:

- 1*1=1
- 1*0=0
- 0*1=0

```
 101
* 11
----
 101
1010
----
1111
```

Note that multiplying by two is extremely easy. To multiply by two, just add a 0 on the end.

## BINARY DIVISION

Follow the same rules as in decimal division. For the sake of simplicity, throw away the remainder.

*For Example*: 111011/11

```
    10011r10
   ---------
11) 111011
    -11
    ---
    101
    -11
    ---
    101
     11
    ---
     10
```

## DECIMAL TO BINARY

Converting from decimal to binary notation is slightly more difficult conceptually, but can easily be done once you know how through the use of algorithms. Begin by thinking of a few examples. We can easily see that the number 3= 2+1. and that this is equivalent to (1*2^1)+(1*2^0).

This translates into putting a "1" in the 2^1 column and a "1" in the 2^0 column, to get "11". Almost as intuitive is the number 5: it is obviously 4+1, which is the same as saying [(2*2) +1], or 2^2+1. This can also be written as [(1*2^2)+(1*2^0)]. Looking at this in columns,

2 ^ 2|2 ^ 2|2 ^ 0<br>
1 0 1<br>
or 101.

What we're doing here is finding the largest power of two within the number (2^2=4 is the largest power of 2 in 5), subtracting that from the number (5-4=1), and finding the largest power of 2 in the remainder (2^0=1 is the largest power of 2 in 1). Then we just put this into columns. This process continues until we have a remainder of 0. Let's take a look at how it works.

*We know that*:

2^0=1<br>
2^1=2<br>
2^2=4<br>
2^3=8<br>
2^4=16<br>
2^5=32<br>
2^6=64<br>
2^7=128

and so on. To convert the decimal number 75 to binary, we would find the largest power of 2 less than 75, which is 64. Thus, we would put a 1 in the 2^6 column, and subtract 64 from 75, giving us 11.

The largest power of 2 in 11 is 8, or 2^3. Put 1 in the 2^3 column, and 0 in 2^4 and 2^5. Subtract 8 from 11 to get 3. Put 1 in the 2^1 column, 0 in 2^2, and subtract 2 from 3. We're left with 1, which goes in 2^0, and we subtract one to get zero. Thus, our number is 1001011.

*Making this algorithm a bit more formal gives us:*

- Let D=number we wish to convert from decimal to binary
- Repeat until D=0
  - Find the largest power of two in D. Let this equal P.
  - Put a 1 in binary column P.
  - Subtract P from D.
- Put zeros in all columns which don't have ones.

This algorithm is a bit awkward. Particularly step 3, "filling in the zeros." Therefore, we should rewrite it such that we ascertain the value of each column individually, putting in 0's and 1's as we go:

- Let D= the number we wish to convert from decimal to binary
- Find P, such that 2^P is the largest power of two smaller than D.
- Repeat until P<0
  - If 2^P<=D then
    a. Put 1 into column P
    b. Subtract 2^P from D
- Else
  - Put 0 into column P
- End if
- Subtract 1 from P

Now that we have an algorithm, we can use it to convert numbers from decimal to binary relatively painlessly. Let's try the number D=55.

- Our first step is to find P. We know that 2^4=16, 2^5=32, and 2^6=64. Therefore, P=5.
- 2^5<=55, so we put a 1 in the 2^5 column: 1———.
- Subtracting 55-32 leaves us with 23. Subtracting 1 from P gives us 4.
- Following step 3 again, 2^4<=23, so we put a 1 in the 2^4 column: 11——.

- Next, subtract 16 from 23, to get 7. Subtract 1 from P gives us 3.
- 2^3>7, so we put a 0 in the 2^3 column: 110—-
- Next, subtract 1 from P, which gives us 2.
- 2^2<=7, so we put a 1 in the 2^2 column: 1101—
- Subtract 4 from 7 to get 3. Subtract 1 from P to get 1.
- 2^1<=3, so we put a 1 in the 2^1 column: 11011-
- Subtract 2 from 3 to get 1. Subtract 1 from P to get 0.
- 2^0<=1, so we put a 1 in the 2^0 column: 110111
- Subtract 1 from 1 to get 0. Subtract 1 from P to get -1.
- P is now less than zero, so we stop.

Another algorithm for converting decimal to binary

However, this is not the only approach possible. We can start at the right, rather than the left.

*All binary numbers are in the form*:

a[n]*2^n + a[n-1]*2^(n-1)+...+a[1]*2^1 + a[0]*2^0

where each a[i] is either a 1 or a 0 (the only possible digits for the binary system). The only way a number can be odd is if it has a 1 in the 2^0 column, because all powers of two greater than 0 are even numbers (2, 4, 8, 16...). This gives us the rightmost digit as a starting point.

Now we need to do the remaining digits. One idea is to "shift" them. It is also easy to see that multiplying and dividing by 2 shifts everything by one column: two in binary is 10, or (1*2^1). Dividing (1*2^1) by 2 gives us (1*2^0), or just a 1 in binary. Similarly, multiplying by 2 shifts in the other direction: (1*2^1)*2=(1*2^2) or 10 in binary. Therefore

{a[n]*2^n + a[n-1]*2^(n-1) +... + a[1]*2^1 + a[0]*2^0}/2

is equal to,

a[n]*2^(n-1) + a[n-1]*2^(n-2) +... + a[1]2^0

Let's look at how this can help us convert from decimal to binary. Take the number 163. We know that since it is odd, there must be a 1 in the 2^0 column (a[0]=1). We also know that it equals 162+1. If we put the 1 in the 2^0 column, we have 162 left, and have to decide how to translate the remaining digits. Two's column: Dividing 162 by 2 gives 81. The number 81 in binary would also have a 1 in the 2^0 column. Since we divided the number by two, we "took out" one power of two. Similarly, the statement a[n-1]*2^(n-1) + a[n-2]*2^(n-2) +... + a[1]*2^0 has a power of two removed. Our "new" 2^0 column now contains a1.

We learned earlier that there is a 1 in the 2^0 column if the number is odd. Since 81 is odd, a[1]=1. Practically, we can simply keep a "running total", which now stands at 11 (a[1]=1 and a[0]=1). Also note that a1 is essentially "remultiplied" by two just by putting it in front of a[0], so it is automatically fit into the correct column.

Four's column: Now we can subtract 1 from 81 to see what remainder we still must place (80). Dividing 80 by 2 gives 40. Therefore, there must be a 0 in the 4's column, (because what we are actually placing is a 2^0 column, and the number is not odd).

Eight's column: We can divide by two again to get 20. This is even, so we put a 0 in the 8's column. Our running total now stands at a[3]=0, a[2]=0, a[1]=1, and a[0]=1.

We can continue in this manner until there is no remainder to place.

*Let's formalize this algorithm*:

- Let D= the number we wish to convert from decimal to binary.
- Repeat until D=0:
  - If D is odd, put "1" in the leftmost open column, and subtract 1 from D.
  - If D is even, put "0" in the leftmost open column.
  - Divide D by 2.

  End Repeat

For the number 163, this works as follows:

- Let D=163
- D is odd, put a 1 in the 2^0 column.
  Subtract 1 from D to get 162.
- Divide D=162 by 2.
  Temporary Result: 01 New D=81
  D does not equal 0, so we repeat step 2.
- D is odd, put a 1 in the 2^1 column.
  Subtract 1 from D to get 80.

- Divide D=80 by 2.
  Temporary Result: 11 New D=40
- D does not equal 0, so we repeat step 2.
- D is even, put a 0 in the 2^2 column.
- Divide D by 2.
  Temporary Result: 011 New D=20
- D is even, put a 0 in the 2^3 column.
- Divide D by 2.
  Temporary Result: 0011 New D=10
- D is even, put a 0 in the 2^4 column.
- Divide D by 2.
  Temporary Result: 00011 New D=5
- D is odd, put a 1 in the 2^5 column.
  Subtract 1 from D to get 4.
- Divide D by 2.
  Temporary Result: 100011 New D=2
- D is even, put a 0 in the 2^6 column.
- Divide D by 2.
  Temporary Result: 0100011 New D=1
- D is odd, put a 1 in the 27 column.
  Subtract 1 from D to get D=0.
- Divide D by 2.
  Temporary Result: 10100011 New D=0

D=0, so we are done, and the decimal number 163 is equivalent to the binary number 10100011.

Since we already knew how to convert from binary to decimal, we can easily verify our result. 10100011 = (1*2^0) + (1*2^1)+(1*2^5)+(1*2^7)=1+2+32+128= 163.

## SIGNED MAGNITUDE

The simplest way to indicate negation is signed magnitude. In signed magnitude, the left-most bit is not actually part of the number, but is just the equivalent of a +/- sign. "0" indicates that the number is positive, "1" indicates negative.

In 8 bits, 00001100 would be 12 (break this down into (1*2^3) + (1*2^2) ). To indicate -12, we would simply put a "1" rather than a "0" as the first bit: 10001100.

## ONE'S COMPLEMENT

In one's complement, positive numbers are represented as usual in regular binary. However, negative numbers are represented differently. To negate a number, replace all zeros with ones, and ones with zeros – flip the bits.

Thus, 12 would be 00001100, and -12 would be 11110011. As in signed magnitude, the leftmost bit indicates the sign (1 is negative, 0 is positive). To compute the value of a negative number, flip the bits and translate as before.

## TWO'S COMPLEMENT

Begin with the number in one's complement. Add 1 if the number is negative. Twelve would be represented as 00001100, and -12 as 11110100. To verify this, let's subtract 1 from 11110100, to get 11110011. If we flip the bits, we get 00001100, or 12 in decimal.

In this notation, "m" indicates the total number of bits. For us (working with 8 bits), it would be excess 2^7. To represent a number (positive or negative) in excess 2^7, begin by taking the number in regular binary representation. Then add 2^7 (=128) to that number. For example, 7 would be 128 + 7=135, or 2^7+2^2+2^1+2^0, and, in binary,10000111. We would represent -7 as 128-7=121, and, in binary, 01111001.

Unless you know which representation has been used, you cannot figure out the value of a number. A number in excess 2^(m-1) is the same as that number in two's complement with the leftmost bit flipped.

To see the advantages and disadvantages of each method, let's try working with them. Using the regular algorithm for binary adition, add (5+12), (-5+12), (-12+-5), and (12+-12) in each system. Then convert back to decimal numbers.

What would the binary number 1011 be in decimal notation?

*Answers*: 1011 = (1*2^3)+(0*2^2)+(1*2^1)+(1*2^0)
= (1*8) + (0*4) + (1*2) + (1*1)
= 11 (in decimal notation)

## Chapter 3

# Number System

### NUMERICAL PRESENTATION

In science, technology, business, and, in fact, most other fields of endeavour, we are constantly dealing with quantities. Quantities are measured, monitored, recorded, manipulated arithmetically, observed, or in some other way utilized in most physical systems. It is important when dealing with various quantities that we be able to represent their values efficiently and accurately. There are basically two ways of representing the numerical value of quantities: analog and digital.

### ANALOG REPRESENTATION

In analog representation a quantity is represented by a voltage, current, or meter movement that is proportional to the value of that quantity. Analog quantities such as those cited above have an important characteristic: they can vary over a continuous range of values.

*Below is a diagram of analog voltage vs time:*

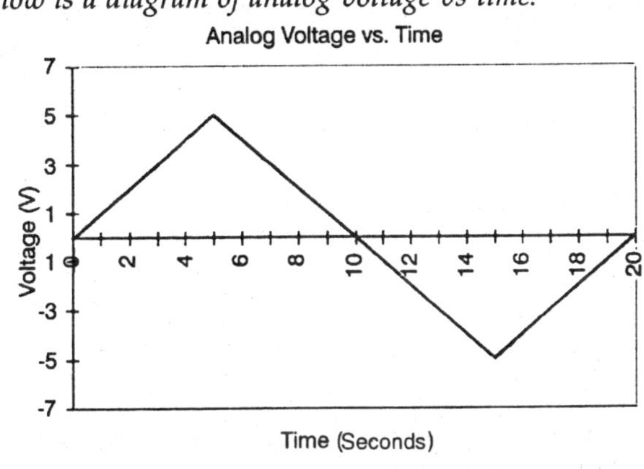

## DIGITAL REPRESENTATION

In digital representation the quantities are represented not by proportional quantities but by symbols called digits. As an example, consider the digital watch, which provides the time of day in the form of decimal digits which represent hours and minutes (and sometimes seconds). As we know, the time of day changes continuously, but the digital watch reading does not change continuously; rather, it changes in steps of one per minute (or per second). In other words, this digital representation of the time of day changes in discrete steps, as compared with the representation of time provided by an analog watch, where the dial reading changes continuously.

*Below is a diagram of digital voltage vs time:*

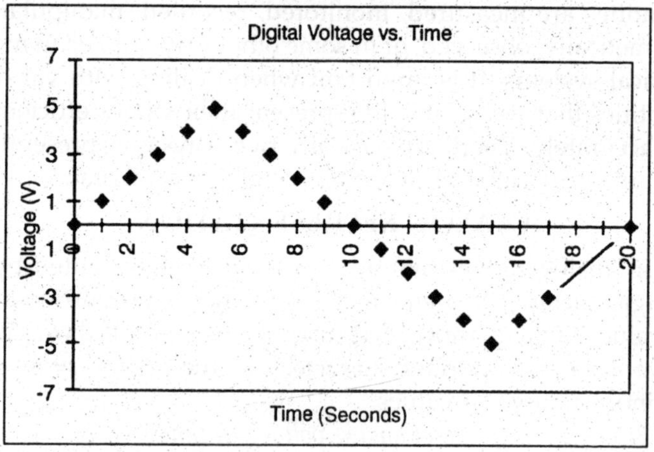


The major difference between analog and digital quantities, then, can be simply stated as follows:

- Analog = Continuous
- Digital = Discrete (step by step)

## ADVANTAGES AND LIMITATIONS

### Advantages

- Easier to design. Exact values of voltage or current are not important, only the range (HIGH or LOW) in which they fall.

- Information storage is easy.
- Accuracy and precision are greater.
- Operation can be programmed. Analog systems can also be programmed, but the variety and complexity of the available operations is severely limited.
- Digital circuits are less affected by noise. As long as the noise is not large enough to prevent us from distinguishing a HIGH from a LOW.
- More digital circuitry can be fabricated on IC chips.

### Limitations

There is really only one major drawback when using digital techniques: The real world is mainly analog. Most physical quantities are analog in nature, and it is these quantities that are often the inputs and outputs that are being monitored, operated on, and controlled by a system.

To take advantage of digital techniques when dealing with analog inputs and outputs, three steps must be follo-wed:

- Convert the real-world analog inputs to digital form. (ADC)
- Process (operate on) the digital information.
- Convert the digital outputs back to real-world analog form. (DAC)

## DIGITAL NUMBER SYSTEM

Many number systems are in use in digital technology. The most common are the decimal, binary, octal, and hexadecimal systems. The decimal system is clearly the most familiar to us because it is a tool that we use every day. Examining some of its characteristics will help us to better understand the other systems.

### DECIMAL SYSTEM

Decimal System The decimal system is composed of 10 numerals or symbols. These 10 symbols are 0, 1, 2, 3, 4, 5, 6, 7, 8, 9; using these symbols as digits of a number, we can express any quantity. The decimal system, also called the base-10 system because it has 10 digits.

| $10^3$ | $10^2$ | $10^1$ | $10^0$ | | $10^{-1}$ | $10^{-2}$ | $10^{-3}$ |
|---|---|---|---|---|---|---|---|
| =1000 | =100 | =10 | =1 | | =0.1 | =0.01 | =0.001 |
| Most Signi- ficant Digit | | | | Decimal point | | Least | Signifi- cant Digit |

## BINARY SYSTEM

In the binary system, there are only two symbols or possible digit values, 0 and 1. This base-2 system can be used to represent any quantity that can be represented in decimal or other number system.

| $2^3$ | $2^2$ | $2^1$ | $2^0$ | | $2^{-1}$ | $2^{-2}$ | $2^{-3}$ |
|---|---|---|---|---|---|---|---|
| =8 | =4 | =2 | =1 | | =1/2 | =1/4 | =1/8 |
| Most Signi- cant Bit | | | | Binary point | | | Least Signifi- cant Bit |

## BINARY COUNTING

**Table.The Binary Counting Sequence.**

| $2^3$=8 | $2^4$=4 | $2^1$=2 | $2^0$=1 | Decimal Equivalent |
|---|---|---|---|---|
| 0 | 0 | 0 | 0 | 0 |
| 0 | 0 | 0 | 1 | 1 |
| 0 | 0 | 1 | 0 | 2 |
| 0 | 0 | 1 | 1 | 3 |
| 0 | 1 | 0 | 0 | 4 |
| 0 | 1 | 0 | 1 | 5 |
| 0 | 1 | 1 | 0 | 6 |
| 0 | 1 | 1 | 1 | 7 |
| 1 | 0 | 0 | 0 | 8 |
| 1 | 0 | 0 | 1 | 9 |
| 1 | 0 | 1 | 0 | 10 |
| 1 | 0 | 1 | 1 | 11 |
| 1 | 1 | 0 | 0 | 12 |
| 1 | 1 | 0 | 1 | 13 |
| 1 | 1 | 1 | 0 | 14 |
| 1 | 1 | 1 | 1 | 15 |

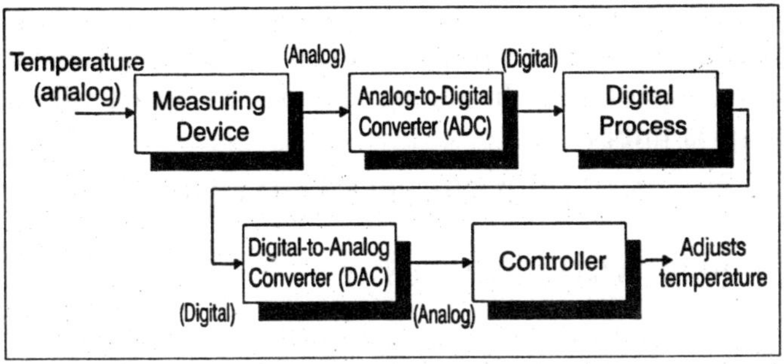


## REPRESENTING BINARY QUANTITIES

In digital systems the information that is being processed is usually presented in binary form. Binary quantities can be represented by any device that has only two operating states or possible conditions.

## EXAMPLE

A switch has only open or closed. We arbitrarily (as we define them) let an open switch represent binary 0 and a closed switch represent binary 1. Thus we can represent any binary number by using series of switches.

## TYPICAL VOLTAGE ASSIGNMENT

*Binary 1*: Any voltage between 2V to 5V

*Binary 0*: Any voltage between 0V to 0.8V

*Not used*: Voltage between 0.8V to 2V, this may cause error in a digital circuit.

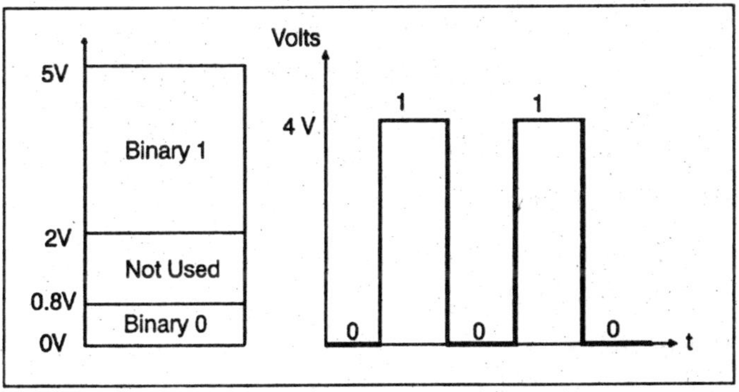


We can see another significant difference between digital

and analog systems. In digital systems, the exact value of a voltage is not important; eg, a voltage of 3.6V means the same as a voltage of 4.3V. In analog systems, the exact value of a voltage is important.

## NUMBER SYSTEM AND CODES

### NUMBERS AND SYMBOLS

The expression of numerical quantities is something we tend to take for granted. This is both a good and a bad thing in the study of electronics.

It is good, in that we're accustomed to the use and manipulation of numbers for the many calculations used in analyzing electronic circuits. On the other hand, the particular system of notation we've been taught from grade school onward is not the system used internally in modern electronic computing devices, and learning any different system of notation requires some re-examination of deeply ingrained assumptions.

First, we have to distinguish the difference between numbers and the symbols we use to represent numbers. A number is a mathematical quantity, usually correlated in electronics to a physical quantity such as voltage, current, or resistance. There are many different types of numbers.

*Here are just a few types, for example:*

**Whole Numbers**

1, 2, 3, 4, 5, 6, 7, 8, 9...

**Integers**

-4, -3, -2, -1, 0, 1, 2, 3, 4...

**Irrational Numbers**

Π (approx. 3.1415927), e (approx. 2.718281828), square root of any prime

**Real Numbers**

(All one-dimensional numerical values, negative and

positive, including zero, whole, integer, and irrational numbers)

## Complex Numbers

3 - j4, 34.5 " 20°

Different types of numbers find different application in the physical world. Whole numbers work well for counting discrete objects, such as the number of resistors in a circuit. Integers are needed when negative equivalents of whole numbers are required. Irrational numbers are numbers that cannot be exactly expressed as the ratio of two integers, and the ratio of a perfect circle's circumference to its diameter (ð) is a good physical example of this.

The non-integer quantities of voltage, current, and resistance that we're used to dealing with in DC circuits can be expressed as real numbers, in either fractional or decimal form.

For AC circuit analysis, however, real numbers fail to capture the dual essence of magnitude and phase angle, and so we turn to the use of complex numbers in either rectangular or polar form.

If we are to use numbers to understand processes in the physical world, make scientific predictions, or balance our checkbooks, we must have a way of symbolically denoting them. In other words, we may know how much money we have in our checking account, but to keep record of it we need to have some system worked out to symbolize that quantity on paper, or in some other kind of form for record-keeping and tracking.

*There are two basic ways we can do this*: analog and digital. With analog representation, the quantity is symbolized in a way that is infinitely divisible. With digital representation, the quantity is symbolized in a way that is discretely packaged. You're probably already familiar with an analog representation of money, and didn't realise it for what it was.

Have you ever seen a fund-raising poster made with a picture of a thermometer on it, where the height of the red column indicated the amount of money collected for the cause?

The more money collected, the taller the column of red ink on the poster.

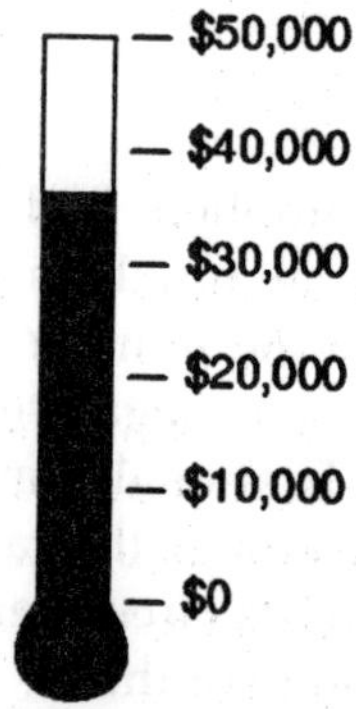


This is an example of an analog representation of a number. There is no real limit to how finely divided the height of that column can be made to symbolize the amount of money in the account. Changing the height of that column is something that can be done without changing the essential nature of what it is. Length is a physical quantity that can be divided as small as you would like, with no practical limit. The slide rule is a mechanical device that uses the very same physical quantity — length — to represent numbers, and to help perform arithmetical operations with two or more numbers at a time. It, too, is an analog device.

On the other hand, a digital representation of that same monetary figure, written with standard symbols (sometimes called ciphers), looks like this: $35,955.38

Unlike the "thermometer" poster with its red column, those symbolic characters above cannot be finely divided: that particular combination of ciphers stand for one quantity and one quantity only. If more money is added to the account (+ $40.12), different symbols must be used to represent the new balance ($35,995.50), or at least the same symbols arranged in different patterns.

This is an example of digital representation. The counterpart to the slide rule (analog) is also a digital device: the abacus, with beads that are moved back and forth on rods to symbolize numerical quantities:

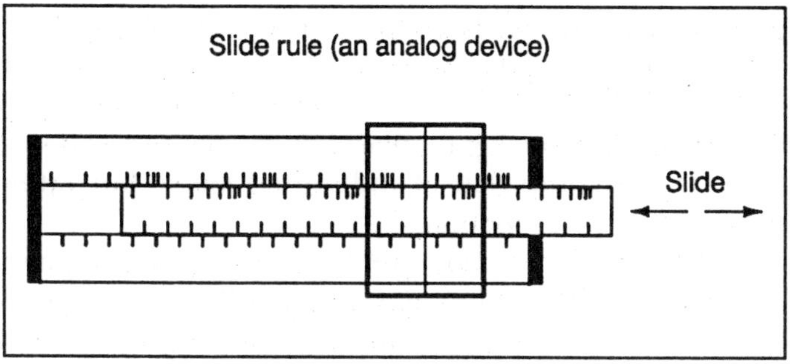


**Fig.** Numerical Quantities are Respresented by the Positioning of the Slide

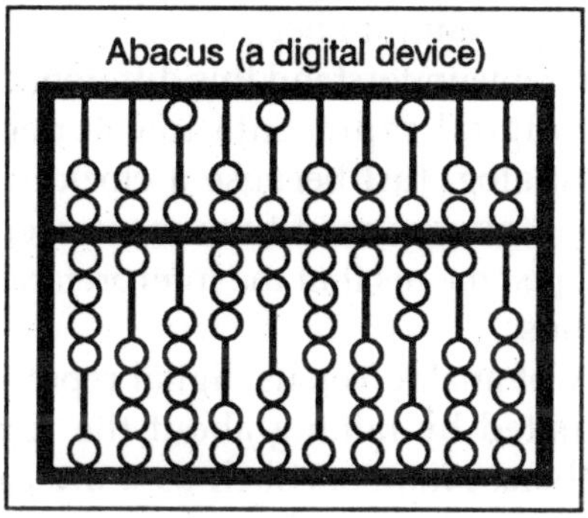


**Fig.** Numerical Quantities are Respresented by the Discrete Positions of the beads.

Let's contrast these two methods of numerical representation:

| Analog | Digital |
|---|---|
| Intuitively understood | Requires training to interpret |
| Infinitely divisible | Discrete |
| Prone to errors of precision | Absolute precision |

Interpretation of numerical symbols is something we tend to take for granted, because it has been taught to us for many years. However, if you were to try to communicate a quantity of something to a person ignorant of decimal numerals, that person could still understand the simple thermometer chart.

The infinitely divisible vs. discrete and precision comparisons are really flip-sides of the same coin. The fact that digital representation is composed of individual, discrete symbols (decimal digits and abacus beads) necessarily means that it will be able to symbolize quantities in precise steps. On the other hand, an analog representation (such as a slide rule's length) is not composed of individual steps, but rather a continuous range of motion.

The ability for a slide rule to characterize a numerical quantity to infinite resolution is a trade-off for imprecision. If a slide rule is bumped, an error will be introduced into the representation of the number that was "entered" into it. However, an abacus must be bumped much harder before its beads are completely dislodged from their places (sufficient to represent a different number).

Please don't misunderstand this difference in precision by thinking that digital representation is necessarily more accurate than analog. Just because a clock is digital doesn't mean that it will always read time more accurately than an analog clock, it just means that the interpretation of its display is less ambiguous.

Divisibility of analog versus digital representation can be further illuminated by talking about the representation of irrational numbers. Numbers such as $\pi$ are called irrational, because they cannot be exactly expressed as the fraction of integers, or whole numbers. Although you might have learned in the past that the fraction 22/7 can be used for $\pi$ in calculations, this is just an approximation. The actual number "pi" cannot be exactly expressed by any finite, or limited, number of decimal places.

*The digits of* $\pi$ *go on forever*:

3. 14159265358979323 84.....

It is possible, at least theoretically, to set a slide rule (or even a thermometer column) so as to perfectly represent the number $\pi$, because analog symbols have no minimum limit to the degree that they can be increased or decreased. If my slide rule shows a figure of 3.141593 instead of 3.141592654, I can bump the slide just a bit more (or less) to get it closer yet. However, with digital representation, such as with an abacus, I would need additional rods (place holders, or digits) to represent π to further degrees of precision. An abacus with 10 rods simply cannot represent any more than 10 digits worth of the number π, no matter how I set the beads.

To perfectly represent π, an abacus would have to have an infinite number of beads and rods! The tradeoff, of course, is the practical limitation to adjusting, and reading, analog symbols. Practically speaking, one cannot read a slide rule's scale to the 10th digit of precision, because the marks on the scale are too coarse and human vision is too limited. An abacus, on the other hand, can be set and read with no interpretational errors at all. Furthermore, analog symbols require some kind of standard by which they can be compared for precise interpretation. Slide rules have markings printed along the length of the slides to translate length into standard quantities. Even the thermometer chart has numerals written along its height to show how much money (in dollars) the red column represents for any given amount of height. Imagine if we all tried to communicate simple numbers to each other by spacing our hands apart varying distances.

The number 1 might be signified by holding our hands 1 inch apart, the number 2 with 2 inches, and so on. If someone held their hands 17 inches apart to represent the number 17, would everyone around them be able to immediately and accurately interpret that distance as 17? Probably not. Some would guess short (15 or 16) and some would guess long (18 or 19). Of course, fishermen who brag about their catches don't mind overestimations in quantity!

Perhaps this is why people have generally settled upon digital symbols for representing numbers, especially whole numbers and integers, which find the most application in everyday life. Using the fingers on our hands, we have a ready means of symbolizing integers from 0 to 10. We can make hash marks on paper, wood, or stone to represent the same quantities quite easily:

**5 + 5 + 3 = 13**

For large numbers, though, the "hash mark" numeration system is too inefficient.

## SYSTEMS OF NUMERATION

The Romans devised a system that was a substantial improvement over hash marks, because it used a variety of symbols (or ciphers) to represent increasingly large quantities. The notation for 1 is the capital letter I. The notation for 5 is the capital letter V.

*Other ciphers possess increasing values*:

X = 10

L = 50

C = 100

D = 500

M = 1000

If a cipher is accompanied by another cipher of equal or lesser value to the immediate right of it, with no ciphers greater than that other cipher to the right of that other cipher, that other cipher's value is added to the total quantity.

Thus, VIII symbolizes the number 8, and CLVII symbolizes the number 157. On the other hand, if a cipher is accompanied by another cipher of lesser value to the immediate left, that other cipher's value is subtracted from the first.

Therefore, IV symbolizes the number 4 (V minus I), and CM symbolizes the number 900 (M minus C). You might have noticed that ending credit sequences for most motion pictures contain a notice for the date of production, in Roman numerals. For the year 1987, it would read: MCMLXXXVII. Let's break this numeral down into its constituent parts, from left to right:

M = 1000

+

CM = 900

+

L = 50

+

XXX = 30

+

V = 5
+
II = 2

Large numbers are very difficult to denote this way, and the left vs. right / subtraction vs. addition of values can be very confusing, too. Another major problem with this system is that there is no provision for representing the number zero or negative numbers, both very important concepts in mathematics.

Roman culture, however, was more pragmatic with respect to mathematics than most, choosing only to develop their numeration system as far as it was necessary for use in daily life.

We owe one of the most important ideas in numeration to the ancient Babylonians, who were the first (as far as we know) to develop the concept of cipher position, or place value, in representing larger numbers. Instead of inventing new ciphers to represent larger numbers, as the Romans did, they re-used the same ciphers, placing them in different positions from right to left. Our own decimal numeration system uses this concept, with only ten ciphers (0, 1, 2, 3, 4, 5, 6, 7, 8, and 9) used in "weighted" positions to represent very large and very small numbers.

Each cipher represents an integer quantity, and each place from right to left in the notation represents a multiplying constant, or weight, for each integer quantity. For example, if we see the decimal notation "1206", we known that this may be broken down into its constituent weight-products as such:

$1206 = 1000 + 200 + 6$

$1206 = (1 \times 1000) + (2 \times 100) + (0 \times 10) + (6 \times 1)$

Each cipher is called a digit in the decimal numeration system, and each weight, or place value, is ten times that of the one to the immediate right. So, we have a ones place, a tens place, a hundreds place, a thousands place, and so on, working from right to left.

Right about now, you're probably wondering why I'm laboring to describe the obvious. Who needs to be told how decimal numeration works, after you've studied math as advanced as algebra and trigonometry? The reason is to better understand other numeration systems, by first knowing the how's and why's of the one you're already used to.

The decimal numeration system uses ten ciphers, and place-weights that are multiples of ten. What if we made a numeration system with the same strategy of weighted places, except with fewer or more ciphers?

The binary numeration system is such a system. Instead of ten different cipher symbols, with each weight constant being ten times the one before it, we only have two cipher symbols, and each weight constant is twice as much as the one before it.

The two allowable cipher symbols for the binary system of numeration are "1" and "0," and these ciphers are arranged right-to-left in doubling values of weight. The rightmost place is the ones place, just as with decimal notation. Proceeding to the left, we have the twos place, the fours place, the eights place, the sixteens place, and so on. For example, the following binary number can be expressed, just like the decimal number 1206, as a sum of each cipher value times its respective weight constant:

$$11010 = 2 + 8 + 16 = 26$$

$$11010 = (1 \times 16) + (1 \times 8) + (0 \times 4) + (1 \times 2) + (0 \times 1)$$

This can get quite confusing, as I've written a number with binary numeration (11010), and then shown its place values and total in standard, decimal numeration form (16 + 8 + 2 = 26). In the above example, we're mixing two different kinds of numerical notation.

To avoid unnecessary confusion, we have to denote which form of numeration we're using when we write (or type!). Typically, this is done in subscript form, with a "2" for binary and a "10" for decimal, so the binary number $11010_2$ is equal to the decimal number $26_{10}$.

The subscripts are not mathematical operation symbols like superscripts (exponents) are. All they do is indicate what system of numeration we're using when we write these symbols for other people to read.

If you see "$3_{10}$", all this means is the number three written using decimal numeration. However, if you see "$3^{10}$", this means something completely different: three to the tenth power (59,049). As usual, if no subscript is shown, the cipher(s) are assumed to be representing a decimal number.

Commonly, the number of cipher types (and therefore, the place-value multiplier) used in a numeration system is called that system's base. Binary is referred to as "base two" numeration, and decimal as "base ten." Additionally, we refer to each cipher position in binary as a bit rather than the familiar word digit used in the decimal system.

Now, why would anyone use binary numeration? The decimal system, with its ten ciphers, makes a lot of sense, being that we have ten fingers on which to count between our two hands. (It is interesting that some ancient central American cultures used numeration systems with a base of twenty. Presumably, they used both fingers and toes to count!!). But the primary reason that the binary numeration system is used in modern electronic computers is because of the ease of representing two cipher states (0 and 1) electronically.

With relatively simple circuitry, we can perform mathematical operations on binary numbers by representing each bit of the numbers by a circuit which is either on (current) or off (no current). Just like the abacus with each rod representing another decimal digit, we simply add more circuits to give us more bits to symbolize larger numbers.

Binary numeration also lends itself well to the storage and retrieval of numerical information: on magnetic tape (spots of iron oxide on the tape either being magnetized for a binary "1" or demagnetized for a binary "0"), optical disks (a laser-burned pit in the aluminum foil representing a binary "1" and an unburned spot representing a binary "0"), or a variety of other media types.

Before we go on to learning exactly how all this is done in digital circuitry, we need to become more familiar with binary and other associated systems of numeration.

## DECIMAL VERSUS BINARY NUMERATION

Let's count from zero to twenty using four different kinds

of numeration systems: hash marks, Roman numerals, decimal, and binary:

| System | Hash Marks | Roman | Decimal | Binary |
|---|---|---|---|---|
| Zero | n/a | n/a | 0 | 0 |
| One | I | I | 1 | 1 |
| Two | II | II | 2 | 10 |
| Three | III | III | 3 | 11 |
| Four | IIII | IV | 4 | 100 |
| Five | /III/ | V | 5 | 101 |
| Six | /III/ I | VI | 6 | 110 |
| Seven | /III/ II | VII | 7 | 111 |
| Eight | /III/ III | VIII | 8 | 1000 |
| Nine | /III/ IIII | IX | 9 | 1001 |
| Ten | /III/ /III/ | X | 10 | 1010 |
| Eleven | /III/ /III/ I | XI | 11 | 1011 |
| Twelve | /III/ /III/ II | XII | 12 | 1100 |
| Thirteen | /III/ /III/ III | XIII | 13 | 1101 |
| Fourteen | /III/ /III/ IIII | XIV | 14 | 1110 |
| Fifteen | /III/ /III/ /III/ | XV | 15 | 1111 |
| Sixteen | /III/ /III/ /III/ /I | XVI | 16 | 10000 |
| Seventeen | /III/ /III/ /III/ II | XVII | 17 | 10001 |
| Eighteen | /III/ /III/ /III/ III | XVIII | 18 | 10010 |
| Nineteen | /III/ /III/ /III/ IIII | XIX | 19 | 10011 |
| Twenty | /III/ /III/ /III/ /III/ | XX | 20 | 10100 |

Neither hash marks nor the Roman system are very practical for symbolizing large numbers. Obviously, place-weighted systems such as decimal and binary are more efficient for the task. Notice, though, how much shorter decimal notation is over binary notation, for the same number of quantities. What takes five bits in binary notation only takes two digits in decimal notation.

This raises an interesting question regarding different numeration systems: how large of a number can be represented with a limited number of cipher positions, or places? With the crude hash-mark system, the number of places IS the largest number that can be represented, since one hash mark "place" is required for every integer step. For place-weighted systems of numeration, however, the answer is found by taking base of the numeration system (10 for decimal, 2 for binary) and raising it to the power of the number of places.

For example, 5 digits in a decimal numeration system can represent 100,000 different integer number values, from 0 to 99,999 (10 to the 5th power = 100,000). 8 bits in a binary numeration system can represent 256 different integer number values, from 0 to 11111111 (binary), or 0 to 255 (decimal), because 2 to the 8th power equals 256. With each additional place position to the number field, the capacity for representing numbers increases by a factor of the base (10 for decimal, 2 for binary).

An interesting footnote for this topic is the one of the first electronic digital computers, the Eniac. The designers of the Eniac chose to represent numbers in decimal form, digitally, using a series of circuits called "ring counters" instead of just going with the binary numeration system, in an effort to minimize the number of circuits required to represent and calculate very large numbers. This approach turned out to be counter-productive, and virtually all digital computers since then have been purely binary in design.

To convert a number in binary numeration to its equivalent in decimal form, all you have to do is calculate the sum of all the products of bits with their respective place-weight constants.

*Convert* $11001101_2$ *to decimal form:*

```
bits           =   1 1 0 0 1 1 0 1.
                   - - - - - - - -
weight         =   1 6 3 1 8 4 2 1
(in decimal        2 4 2 6
notation)          8
```

The bit on the far right side is called the Least Significant Bit (LSB), because it stands in the place of the lowest weight (the one's place). The bit on the far left side is called the Most Significant Bit (MSB), because it stands in the place of the highest weight (the one hundred twenty-eight's place). Remember, a bit value of "1" means that the respective place weight gets added to the total value, and a bit value of "0" means that the respective place weight does not get added to the total value. With the above example, *we have*:

$$128_{10} + 64_{10} + 8_{10} + 4_{10} + 1_{10} = 205_{10}$$

If we encounter a binary number with a dot (.), called a "binary point" instead of a decimal point, we follow the same procedure, realizing that each place weight to the right of the point is one-half the value of the one to the left of it (just as each place weight to the right of a decimal point is one-tenth the weight of the one to the left of it).

*For example*:

*Convert 101.011$_2$ to decimal form*:

```
.
bits    =        1 0 1. 0 1 1
                 - - - - - - -
.
weight  =        4 2 1  1 1 1
(in decimal             / / /
notation)               2 4 8
```

$$4_{10} + 1_{10} + 0.25_{10} + 0.125_{10} = 5.375_{10}$$

## OCTAL AND HEXADECIMAL NUMERATION

Because binary numeration requires so many bits to represent relatively small numbers compared to the economy of the decimal system, analyzing the numerical states inside of digital electronic circuitry can be a tedious task. Computer programmers who design sequences of number codes instructing a computer what to do would have a very difficult task if they were forced to work with nothing but long strings of 1's and 0's, the "native language" of any digital circuit.

To make it easier for human engineers, technicians, and programmers to "speak" this language of the digital world, other systems of place-weighted numeration have been made which are very easy to convert to and from binary. One of those numeration systems is called octal, because it is a place-weighted system with a base of eight. Valid ciphers include the symbols 0, 1, 2, 3, 4, 5, 6, and 7. Each place weight differs from the one next to it by a factor of eight.

Another system is called hexadecimal, because it is a place-weighted system with a base of sixteen. Valid ciphers include the normal decimal symbols 0, 1, 2, 3, 4, 5, 6, 7, 8, and 9, plus six alphabetical characters A, B, C, D, E, and F, to make a total of sixteen. As you might have guessed already, each place weight differs from the one before it by a factor of sixteen.

Let's count again from zero to twenty using decimal, binary, octal, and hexadecimal to contrast these systems of numeration:

| Number | Decimal | Binary | Octal | Hexadecimal |
|---|---|---|---|---|
| Zero | 0 | 0 | 0 | 0 |
| One | 1 | 1 | 1 | 1 |
| Two | 2 | 10 | 2 | 2 |
| Three | 3 | 11 | 3 | 3 |
| Four | 4 | 100 | 4 | 4 |
| Five | 5 | 101 | 5 | 5 |
| Six | 6 | 110 | 6 | 6 |
| Seven | 7 | 111 | 7 | 7 |
| Eight | 8 | 1000 | 10 | 8 |
| Nine | 9 | 1001 | 11 | 9 |
| Ten | 10 | 1010 | 12 | A |
| Eleven | 11 | 1011 | 13 | B |
| Twelve | 12 | 1100 | 14 | C |
| Thirteen | 13 | 1101 | 15 | D |
| Fourteen | 14 | 1110 | 16 | E |
| Fifteen | 15 | 1111 | 17 | F |
| Sixteen | 16 | 10000 | 20 | 10 |
| Seventeen | 17 | 10001 | 21 | 11 |
| Eighteen | 18 | 10010 | 22 | 12 |
| Nineteen | 19 | 10011 | 23 | 13 |
| Twenty | 20 | 10100 | 24 | 14 |

Octal and hexadecimal numeration systems would be pointless if not for their ability to be easily converted to and from binary notation. Their primary purpose in being is to serve as a "shorthand" method of denoting a number represented electronically in binary form.

Because the bases of octal (eight) and hexadecimal (sixteen) are even multiples of binary's base (two), binary bits can be grouped together and directly converted to or from their respective octal or hexadecimal digits. With octal, the binary bits are grouped in three's (because $2^3 = 8$), and with hexadecimal, the binary bits are grouped in four's (because $2^4 = 16$):

## Binary to Octal Conversion

*Convert $10110111.1_2$ to octal:*

```
                     implied zero       implied zeros
                            |                  ||
                            010  110 111  100
Convert each group of       ### ### ###. ###
bits
to its octal equivalent: 2    6   7  4
```

*Answer:* $10110111.1_2 = 267.4_8$

We had to group the bits in three's, from the binary point left, and from the binary point right, adding (implied) zeros as necessary to make complete 3-bit groups. Each octal digit was translated from the 3-bit binary groups. Binary-to-Hexadecimal conversion is much the same:

## Binary to Hexadecimal Conversion

*Convert $10110111.1_2$ to hexadecimal:*

```
                                 implied zeros
                                     |||
                            1011 0111 1000
Convert each group          - - - -. - -
of bits to its
hexadecimal equivalent:  B     7    8
```

*Answer:* $10110111.1_2 = B7.8_{16}$

Here we had to group the bits in four's, from the binary point left, and from the binary point right, adding (implied) zeros as necessary to make complete 4 bit groups:

Likewise, the conversion from either octal or hexadecimal to binary is done by taking each octal or hexadecimal digit and converting it to its equivalent binary (3 or 4 bit) group, then putting all the binary bit groups together.

Incidentally, hexadecimal notation is more popular, because binary bit groupings in digital equipment are commonly multiples of eight (8, 16, 32, 64, and 128 bit), which are also multiples of 4. Octal, being based on binary bit groups of 3, doesn't work out evenly with those common bit group sizings.

## OCTAL AND HEXADECIMAL TO DECIMAL CONVERSION

Although the prime intent of octal and hexadecimal numeration systems is for the "shorthand" representation of binary numbers in digital electronics, we sometimes have the need to convert from either of those systems to decimal form. Of course, we could simply convert the hexadecimal or octal format to binary, then convert from binary to decimal, since we already know how to do both, but we can also convert directly.

Because octal is a base-eight numeration system, each place-weight value differs from either adjacent place by a factor of eight. For example, the octal number 245.37 can be broken down into place values as such:

```
octal
digits =2     4     5     .     3     7
.       -     -     -     -     -     -
weight =      6     8     1           1     1
(in deci-     4                       /     /
mal notation)                   8     6
.                                     4
```

The decimal value of each octal place-weight times its respective cipher multiplier can be determined as follows:

$$(2 \times 64_{10}) + (4 \times 8_{10}) + (5 \times 1_{10}) + (3 \times 0.125_{10}) + (7 \times 0.015625_{10}) = 165.484375_{10}$$

The technique for converting hexadecimal notation to decimal is the same, except that each successive place-weight changes by a factor of sixteen. Simply denote each digit's weight, multiply each hexadecimal digit value by its respective weight (in decimal form), then add up all the decimal values to get a total. For example, the hexadecimal number $30F.A9_{16}$ can be converted like this:

```
hexadecimal
digits =    3  0  F  .  A  9
```

```
.               -   -   -       -   -
weight =        2   1   1       1   1
(in decimal     5   6           /   /
notation)       6               1   2
.                               6   5
.                                   6
```

$$(3 \times 256_{10}) + (0 \times 16_{10}) + (15 \times 1_{10}) + (10 \times 0.0625_{10}) + (9 \times 0.00390625_{10}) = 783.66015625_{10}$$

These basic techniques may be used to convert a numerical notation of any base into decimal form, if you know the value of that numeration system's base.

## CONVERSION FROM DECIMAL NUMERATION

Because octal and hexadecimal numeration systems have bases that are multiples of binary (base 2), conversion back and forth between either hexadecimal or octal and binary is very easy. Also, because we are so familiar with the decimal system, converting binary, octal, or hexadecimal to decimal form is relatively easy (simply add up the products of cipher values and place-weights). However, conversion from decimal to any of these "strange" numeration systems is a different matter.

The method which will probably make the most sense is the "trial-and-fit" method, where you try to "fit" the binary, octal, or hexadecimal notation to the desired value as represented in decimal form. For example, let's say that I wanted to represent the decimal value of 87 in binary form. Let's start by drawing a binary number field, complete with place-weight values:.

```
.               -   -   -   -   -   -   -   -
weight=         1   6   3   1   8   4   2   1
(in decimal     2   4   2   6
notation)       8
```

Well, we know that we won't have a "1" bit in the 128's place, because that would immediately give us a value greater than 87. However, since the next weight to the right (64) is less than 87, we know that we must have a "1" there.

```
                1
```

```
.       -   -   -   -   -   -   - Decimal
                                  value so
                              far = 64₁₀
weight =6    3   1   8   4   2   1
(in deci-    4   2   6
mal notation)
```

If we were to make the next place to the right a "1" as well, our total value would be $64_{10} + 32_{10}$, or $96_{10}$. This is greater than $87_{10}$, so we know that this bit must be a "0". If we make the next (16's) place bit equal to "1," this brings our total value to $64_{10} + 16_{10}$, or $80_{10}$, which is closer to our desired value ($87_{10}$) without exceeding it:

```
.             1   0   1
.             -   -   -   -   -   -   - Decimal
                                        value so
                                           far = 80₁₀
weight =      6   3   1   8   4   2   1
(in decimal   4   2   6
notation)
```

By continuing in this progression, setting each lesser-weight bit as we need to come up to our desired total value without exceeding it, we will eventually arrive at the correct figure:

```
.             1   0   1   0   1   1   1
.             -   -   -   -   -   -   - Decimal
                                        value so
                                    far = 87₁₀
weight =      6   3   1   8   4   2   1
(in decimal   4   2   6
notation)
```

This trial-and-fit strategy will work with octal and hexadecimal conversions, too. Let's take the same decimal figure, $87_{10}$, and convert it to octal numeration:.

```
              -   -   -
.
weight =      6   8   1
(in decimal   4
notation)
```

If we put a cipher of "1" in the 64's place, we would have

a total value of $64_{10}$ (less than $87_{10}$). If we put a cipher of "2" in the 64's place, we would have a total value of $128_{10}$ (greater than $87_{10}$). This tells us that our octal numeration must start with a "1" in the 64's place:

```
.              1
.              -   -   - Decimal value so far = 64₁₀
weight =       6   8   1
(in decimal 4
notation)
```

Now, we need to experiment with cipher values in the 8's place to try and get a total (decimal) value as close to 87 as possible without exceeding it. Trying the first few cipher options, we get:

$$"1" = 64_{10} + 8_{10} = 72_{10}$$
$$"2" = 64_{10} + 16_{10} = 80_{10}$$
$$"3" = 64_{10} + 24_{10} = 88_{10}$$

A cipher value of "3" in the 8's place would put us over the desired total of $87_{10}$, so "2" it is!

```
.              1   2
.              -   -   - Decimal value so far = 80₁₀
weight =       6   8   1
(in decimal 4
notation)
```

Now, all we need to make a total of 87 is a cipher of "7" in the 1's place:

```
.              1   2   7
.              -   -   - Decimal value so far = 87₁₀
weight =       6   8   1
(in decimal 4
notation)
```

Of course, if you were paying attention during the last section on octal/binary conversions, you will realise that we can take the binary representation of (decimal) $87_{10}$, which we previously determined to be $1010111_2$, and easily convert from that to octal to check our work:

```
.    Implied zeros
.        ||
.        001 010 111 Binary
```

```
.        _   _   _
.        1   2   7   Octal
.
```

*Answer*: $1010111_2 = 127_8$

Can we do decimal-to-hexadecimal conversion the same way? Sure, but who would want to? This method is simple to understand, but laborious to carry out. There is another way to do these conversions, which is essentially the same (mathematically), but easier to accomplish.

This other method uses repeated cycles of division (using decimal notation) to break the decimal numeration down into multiples of binary, octal, or hexadecimal place-weight values. In the first cycle of division, we take the original decimal number and divide it by the base of the numeration system that we're converting to (binary=2 octal=8, hex=16).

Then, we take the whole-number portion of division result (quotient) and divide it by the base value again, and so on, until we end up with a quotient of less than 1. The binary, octal, or hexadecimal digits are determined by the "remainders" left over by each division step. Let's see how this works for binary, with the decimal example of $87_{10}$:

```
. 87         Divide 87 by 2, to get a quotient
        of 43.5
. — =        43.5 Division "remainder" = 1, or
        the < 1 portion
. 2          of the quotient times the divisor
        (0.5 × 2)
.
. 43         Take the whole-number portion of 43.5
             (43)
. — = 21.5   and divide it by 2 to get 21.5, or
             21 with
. 2          a remainder of 1
.
. 21         And so on... remainder = 1 (0.5
        x 2)
. — = 10.5
. 2
```

```
.
. 10          And so on... remainder = 0
. — = 5.0
. 2
.
. 5           And so on... remainder = 1 (0.5
         x 2)
. — = 2.5
. 2
.
. 2           And so on... remainder = 0
. — = 1.0
. 2
.
. 1           ... until we get a quotient of
     less than 1
. — = 0.5    remainder = 1 (0.5 × 2)
. 2
```

The binary bits are assembled from the remainders of the successive division steps, beginning with the LSB and proceeding to the MSB. In this case, we arrive at a binary notation of $1010111_2$. When we divide by 2, we will always get a quotient ending with either ".0" or ".5", i.e. a remainder of either 0 or 1. As was said before, this repeat-division technique for conversion will work for numeration systems other than binary. If we were to perform successive divisions using a different number, such as 8 for conversion to octal, we will necessarily get remainders between 0 and 7. Let's try this with the same decimal number, $87_{10}$:

```
. 87                  Divide 87 by 8, to get a
                quotient of 10.875
. — =                 10.875 Division "remainder"
                      = 7, or the < 1 portion
. 8                   of the quotient times the
                divisor(.875 × 8)
.
. 10
. — = 1.25            remainder = 2
```

```
. 8
.
. 1
. — = 0.125         Quotient is less than 1, so
                    we'll stop here.
. 8 Remainder =     1
.
. Result: 87₁₀ =    127₈
```

We can use a similar technique for converting numeration systems dealing with quantities less than 1, as well. For converting a decimal number less than 1 into binary, octal, or hexadecimal, we use repeated multiplication, taking the integer portion of the product in each step as the next digit of our converted number. Let's use the decimal number $0.8125_{10}$ as an example, converting to binary:

```
. 0.8125 × 2 = 1.625   Integer portion of
                    product = 1
.
. 0.625 × 2 = 1.25       Take < 1 portion of
                    product and remultiply
.                        Integer portion of
                    product = 1
.
. 0.25 × 2 = 0.5         Integer portion of
                    product = 0
.
. 0.5 × 2 = 1.0          Integer portion of
                    product = 1
.                        Stop when product is a
                    pure integer
.                        (ends with.0)
.
. Result: 0.8125₁₀  =    0.1101₂
```

As with the repeat-division process for integers, each step gives us the next digit (or bit) further away from the "point." With integer (division), we worked from the LSB to the MSB (right-to-left), but with repeated multiplication, we worked from the left to the right. To convert a decimal number greater than 1, with a < 1 component, we must use both techniques, one at a time. Take the decimal example of $54.40625_{10}$, converting to binary:

*Repeated division for the integer portion:*

- $\frac{54}{2} = 27.0$ Remainder = 0
- $\frac{27}{2} = 13.5$ Remainder = 1 (0.5 × 2)
- $\frac{13}{2} = 6.5$ Remainder = 1 (0.5 × 2)
- $\frac{6}{2} = 3.0$ Remainder = 0
- $\frac{3}{2} = 1.5$ Remainder = 1 (0.5 × 2)
- $\frac{1}{2} = 0.5$ Remainder = 1 (0.5 × 2)

*Partial answer*: $54_{10} = 110110_2$

Repeated multiplication for the < 1 portion:

- 0.40625 × 2 = 0.8125 Integer portion of product = 0
- 0.8125 × 2 = 1.625 Integer portion of product = 1

. 0.625 × 2 = 1.25 Integer portion of product = 1

.

. 0.25 × 2 = 0.5 Integer portion of product = 0

.

. 0.5 × 2 = 1.0 Integer portion of product = 1

.

. Partial answer: $0.40625_{10} = 0.01101_2$

.

. Complete answer: $54_{10} + 0.40625_{10} = 54.40625_{10}$

.

. $110110_2 + 0.01101_2 = 110110.01101_2$

## BINARY-TO-DECIMAL CONVERSION

Any binary number can be converted to its decimal equivalent simply by summing together the weights of the various positions in the binary number which contain a 1.

$1\,1\,0\,1\,1_2$ (binary)

$2^4 + 2^3 + 0 + 2^1 + 2^0 = 16 + 8 + 0 + 2 + 1$

$= 27_{10}$ (decimal)

and,

$1\,0\,1\,1\,0\,1\,0\,1_2$ (binary)

$2^7 + 0 + 2^5 + 2^4 + 0 + 2^2 + 0 + 2^0 = 128 + 0 + 32 + 16 + 0 + 4 + 0 + 1$

$= 181_{10}$ (decimal)

You should noticed the method is find the weights (i.e., powers of 2) for each bit position that contains a 1, and then to add them up.

## DECIMAL-TO-BINARY CONVERSION

*There are 2 methods:*

(A) Revese of Binary-To-Digital Method

$45_{10} = 32 + 0 + 8 + 4 + 0 + 1$

$= 2^5+0+2^3+2^2+0+2^0$

$= 1\,0\,1\,1\,0\,1_2$

(B) Repeat Division

This method uses repeated division by 2. Eg. convert $25_{10}$ to binary.

| | | |
|---|---|---|
| 25/ 2 | = 12+ remainder of 1 | 1(Least Significant Bit) |
| 12/ 2 | = 6 + remainder of 0 | 0 |
| 6 / 2 | = 3 + remainder of 0 | 0 |
| 3 / 2 | = 1 + remainder of 1 | 1 |
| 1 / 2 | = 0 + remainder of 1 | 1(Most Significant Bit) |
| Result | $25_{10}$ = | $1\,1\,0\,0\,1_2$ |

*The Flow chart for repeated-division method is as follow:*

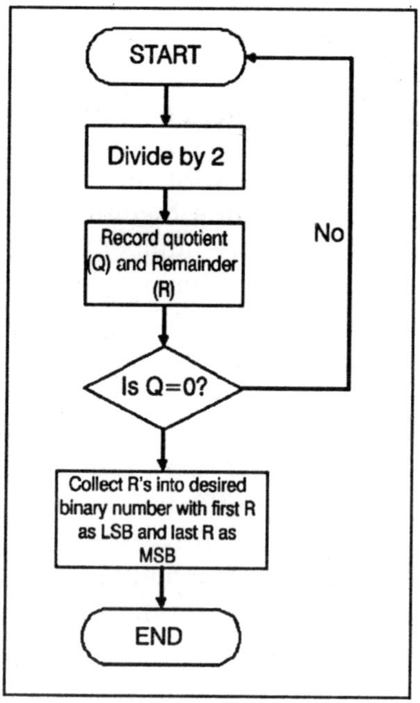


## OCTAL NUMBER SYSTEM

The octal number system has a base of eight, meaning that it has eight possible digits: 0,1,2,3,4,5,6,7.

| $8^3$ | $8^2$ | $8^1$ | $8^0$ | | $8^{-1}$ | $8^{-2}$ | $8^{-3}$ |
|---|---|---|---|---|---|---|---|
| =512 | =64 | =8 | =1 | . | =1/8 | =1/64 | =1/512 |
| Most Significant Digit | | | | Octal point | | | Least Significant Digit |

## OCTAL TO DECIMAL CONVERSION

eg. $24.6_8 = 2 \times (8^1) + 4 \times (8^0) + 6 \times (8^{-1}) = 20.75_{10}$

### Binary-To-Octal / Octal-To-Binary Conversion

| Octal Digit | 0 | 1 | 2 | 3 | 4 | 5 | 6 | 7 |
|---|---|---|---|---|---|---|---|---|
| Binary | 000 | 001 | 010 | 011 | 100 | 101 | 110 | 111 |

*Equivalent*:

Each Octal digit is represented by three bits of binary digit.

eg. $100\ 111\ 010_2 = (100)\ (111)\ (010)_2 = 4\ 7\ 2_8$

## REPEAT DIVISION

This method uses repeated division by 8. Eg. convert $177_{10}$ to octal and binary:

177/8 = 22+ remainder of 1 1 (Least Significant Bit)
22/ 8 = 2 + remainder of 6 6
2 / 8 = 0 + remainder of 2 2 (Most Significant Bit)
Result$177_{10}$ = $261_8$
Convert to Binary = $010110001_2$

## DECIMAL-TO-BINARY CONVERSION

*There are 2 methods*:

(A) Revese of Binary-To-Digital Method

$45_{10} = 32 + 0 + 8 + 4 + 0 + 1$
$= 2^5+0+2^3+2^2+0+2^0$
$= 1\ 0\ 1\ 1\ 0\ 1_2$

(B) Repeat Division

This method uses repeated division by 2. Eg. convert $25_{10}$ to binary,

25/ 2 = 12+ remainder of 1 1 (Least Significant Bit)
12/ 2 = 6 + remainder of 0 0
6 / 2 = 3 + remainder of 0 0
3 / 2 = 1 + remainder of 1 1
1 / 2 = 0 + remainder of 1 1 (Most Significant Bit)
Result $25_{10}$ = $1\ 1\ 0\ 0\ 1_2$

*The Flow chart for repeated-division method is as follow*:

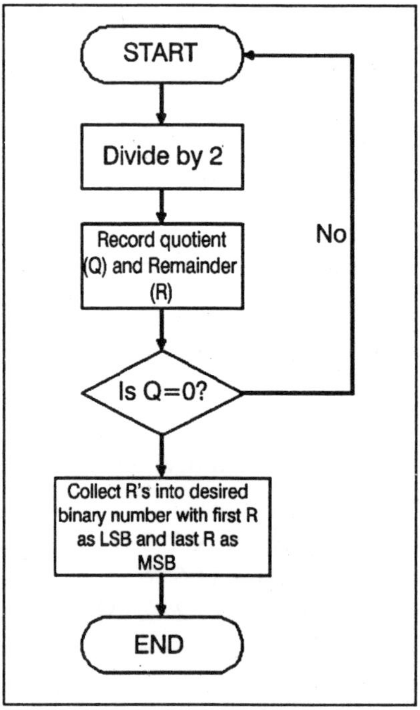
START
Divide by 2
Record quotient (Q) and Remainder (R)
No
Is Q=0?
Collect R's into desired binary number with first R as LSB and last R as MSB
END

# Chapter 4

# Logic Gates

## INTRODUCTION

### THE NOT GATE

The single-transistor inverter circuit illustrated earlier is actually too crude to be of practical use as a gate. Real inverter circuits contain more than one transistor to maximize voltage gain (so as to ensure that the final output transistor is either in full cutoff or full saturation), and other components designed to reduce the chance of accidental damage.

Shown here is a schematic diagram for a real inverter circuit, complete with all necessary components for efficient and reliable operation:

*Practical inverter (NOT) circuit*

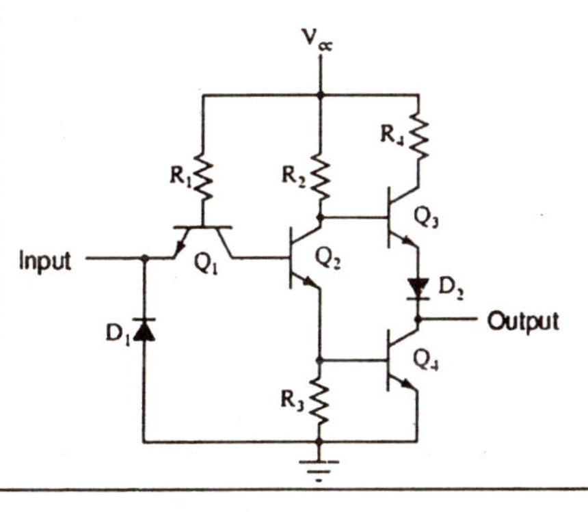


This circuit is composed exclusively of resistors and bipolar transistors. Bear in mind that other circuit designs are capable of performing the NOT gate function, including designs substituting field-effect transistors for bipolar (discussed later in this chapter). Let's analyse this circuit for the condition where the input is "high," or in a binary "1" state. We can simulate this by showing the input terminal connected to $V_{cc}$ through a switch:

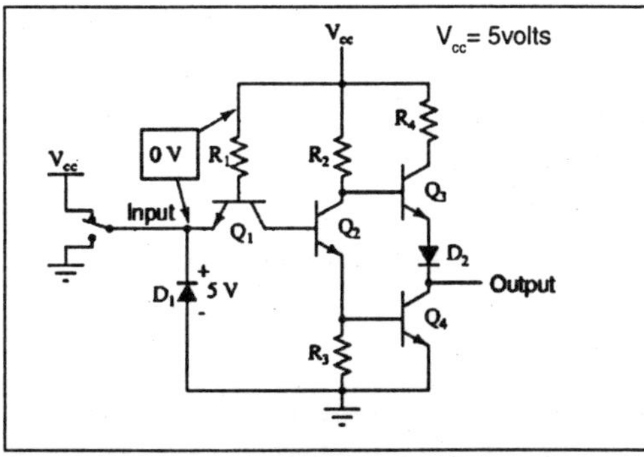


In this case, diode $D_1$ will be reverse-biased, and therefore not conduct any current. In fact, the only purpose for having $D_1$ in the circuit is to prevent transistor damage in the case of a negative voltage being impressed on the input (a voltage that is negative, rather than positive, with respect to ground). With no voltage between the base and emitter of transistor $Q_1$, we would expect no current through it, either. However, as strange as it may seem, transistor $Q_1$ is not being used as is customary for a transistor. In reality, $Q_1$ is being used in this circuit as nothing more than a back-to-back pair of diodes. The following schematic shows the real function of $Q_1$:

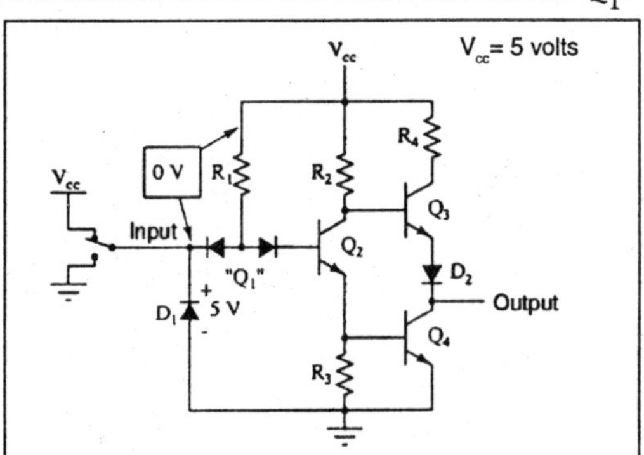

The purpose of these diodes is to "steer" current to or away from the base of transistor $Q_2$, depending on the logic level of the input. Exactly how these two diodes are able to "steer" current isn't exactly obvious at first inspection, so a short example may be necessary for understanding.

Suppose we had the following diode/resistor circuit, representing the base-emitter junctions of transistors $Q_2$ and $Q_4$ as single diodes, stripping away all other portions of the circuit so that we can concentrate on the current "steered" through the two back-to-back diodes:

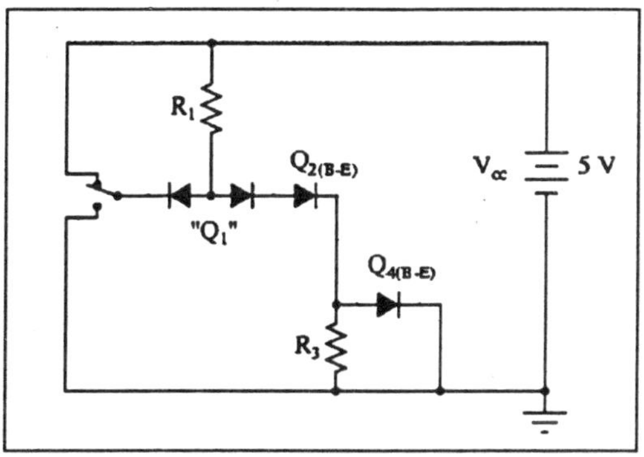


With the input switch in the "up" position (connected to $V_{cc}$), it should be obvious that there will be no current through the left steering diode of $Q_1$, because there isn't any voltage in the switch-diode-$R_1$-switch loop to motivate electrons to flow. However, there will be current through the right steering diode of $Q_1$, as well as through $Q_2$'s base-emitter diode junction and $Q_4$'s base-emitter diode junction:

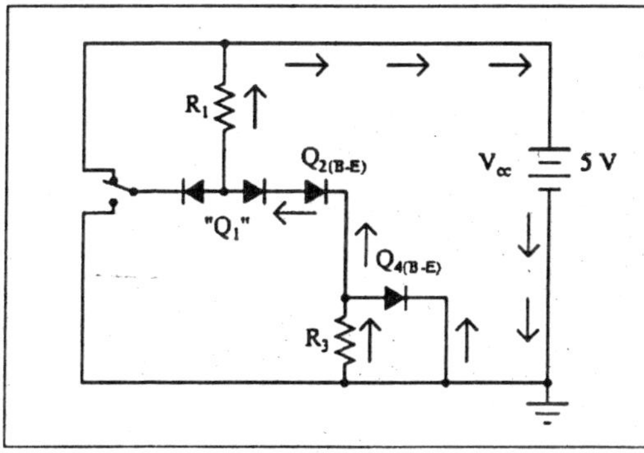

This tells us that in the real gate circuit, transistors $Q_2$ and $Q_4$ will have base current, which will turn them on to conduct collector current.

The total voltage dropped between the base of $Q_1$ (the node joining the two back-to-back steering diodes) and ground will be about 2.1 volts, equal to the combined voltage drops of three PN junctions: the right steering diode, $Q_2$'s base-emitter diode, and $Q_4$'s base-emitter diode.

Now, let's move the input switch to the "down" position and see what happens:

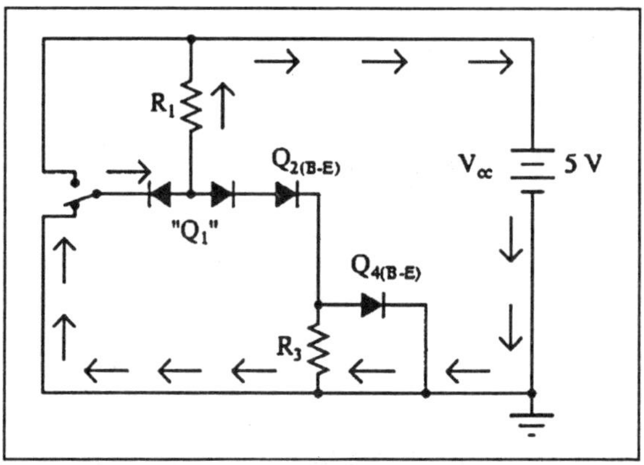


If we were to measure current in this circuit, we would find that all of the current goes through the left steering diode of $Q_1$ and none of it through the right diode.

Why is this? It still appears as though there is a complete path for current through $Q_4$'s diode, $Q_2$'s diode, the right diode of the pair, and $R_1$, so why will there be no current through that path?

Remember that PN junction diodes are very nonlinear devices: they do not even begin to conduct current until the forward voltage applied across them reaches a certain minimum quantity, approximately 0.7 volts for silicon and 0.3 volts for germanium. And then when they begin to conduct current, they will not drop substantially more than 0.7 volts. When the switch in this circuit is in the "down" position, the left diode of the steering diode pair is fully conducting, and so it drops about 0.7 volts across it and no more.

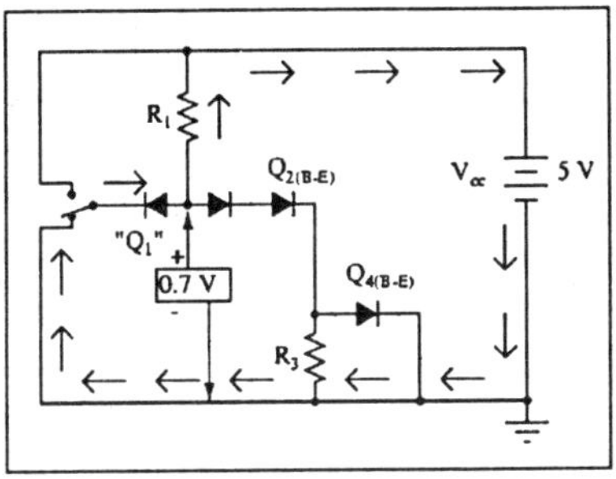


Recall that with the switch in the "up" position (transistors $Q_2$ and $Q_4$ conducting), there was about 2.1 volts dropped between those same two points ($Q_1$'s base and ground), which also happens to be the minimum voltage necessary to forward-bias three series-connected silicon PN junctions into a state of conduction. The 0.7 volts provided by the left diode's forward voltage drop is simply insufficient to allow any electron flow through the series string of the right diode, $Q_2$'s diode, and the $R_3//Q_4$ diode parallel subcircuit, and so no electrons flow through that path. With no current through the bases of either transistor $Q_2$ or $Q_4$, neither one will be able to conduct collector current: transistors $Q_2$ and $Q_4$ will both be in a state of cutoff. Consequently, this circuit configuration allows 100 per cent switching of $Q_2$ base current (and therefore control over the rest of the gate circuit, including voltage at the output) by diversion of current through the left steering diode.

In the case of our example gate circuit, the input is held "high" by the switch (connected to $V_{cc}$), making the left steering diode (zero voltage dropped across it). However, the right steering diode is conducting current through the base of $Q_2$, through resistor $R_1$:

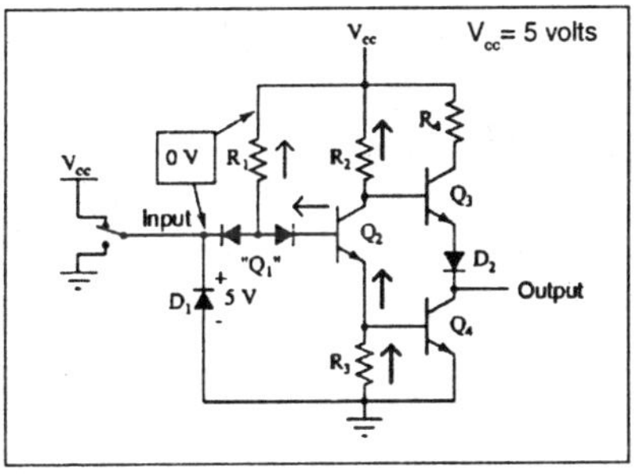

With base current provided, transistor $Q_2$ will be turned "on." More specifically, it will be saturated by virtue of the more-than-adequate current allowed by $R_1$ through the base. With $Q_2$ saturated, resistor $R_3$ will be dropping enough voltage to forward-bias the base-emitter junction of transistor $Q_4$, thus saturating it as well:

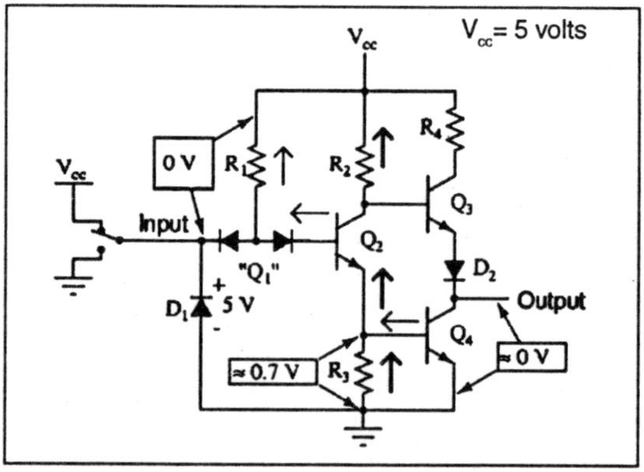


With $Q_4$ saturated, the output terminal will be almost directly shorted to ground, leaving the output terminal at a voltage (in reference to ground) of almost 0 volts, or a binary "0" ("low") logic level. Due to the presence of diode $D_2$, there will not be enough voltage between the base of $Q_3$ and its emitter to turn it on, so it remains in cutoff.

Let's see now what happens if we reverse the input's logic level to a binary "0" by actuating the input switch:

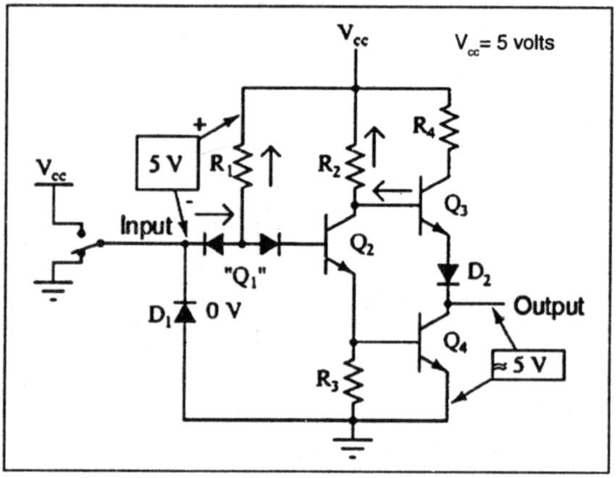

Now there will be current through the left steering diode of $Q_1$ and no current through the right steering diode. This eliminates current through the base of $Q_2$, thus turning it off. With $Q_2$ off, there is no longer a path for $Q_4$ base current, so $Q_4$ goes into cutoff as well.

$Q_3$, on the other hand, now has sufficient voltage dropped between its base and ground to forward-bias its base-emitter junction and saturate it, thus raising the output terminal voltage to a "high" state. In actuality, the output voltage will be somewhere around 4 volts depending on the degree of saturation and any load current, but still high enough to be considered a "high" (1) logic level.

With this, our simulation of the inverter circuit is complete: a "1" in gives a "0" out, and visa-versa.

The astute observer will note that this inverter circuit's input will assume a "high" state of left floating (not connected to either $V_{cc}$ or ground). With the input terminal left unconnected, there will be no current through the left steering diode of $Q_1$, leaving all of $R_1$'s current to go through $Q_2$'s base, thus saturating $Q_2$ and driving the circuit output to a "low" state:

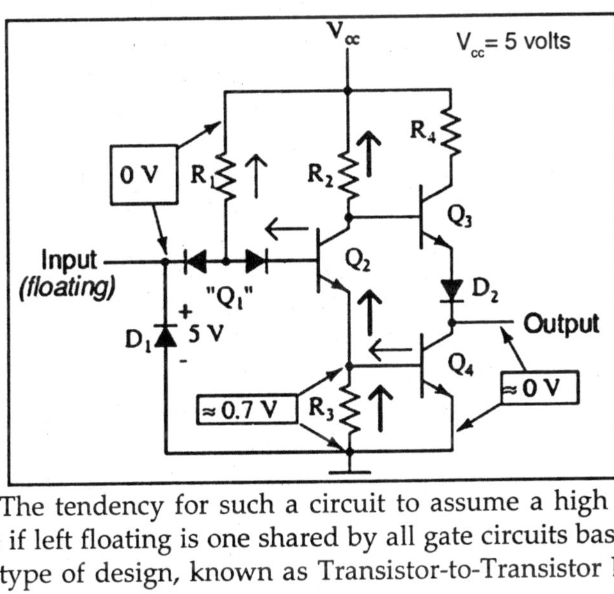


The tendency for such a circuit to assume a high input state if left floating is one shared by all gate circuits based on this type of design, known as Transistor-to-Transistor Logic, or TTL. This characteristic may be taken advantage of in simplifying the design of a gate's output circuitry, knowing that the outputs of gates typically drive the inputs of other gates. If the input of a TTL gate circuit assumes a high state when floating, then the output of any gate driving a TTL input need only provide a path to ground for a low state and be floating for a high state. This concept may require further elaboration for full understanding, so I will explore it in detail here.

A gate circuit as we have just analysed has the ability to handle output current in two directions: in and out. Technically, this is known as sourcing and sinking current, respectively.

When the gate output is high, there is continuity from the output terminal to $V_{cc}$ through the top output transistor ($Q_3$), allowing electrons to flow from ground, through a load, into the gate's output terminal, through the emitter of $Q_3$, and eventually up to the $V_{cc}$ power terminal (positive side of the DC power supply):

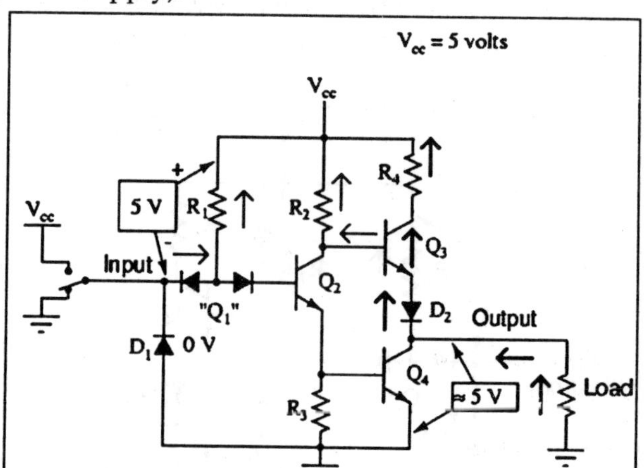


**Fig.** Inventer Gate Sourcing Current.

To simplify this concept, we may show the output of a gate circuit as being a double-throw switch, capable of connecting the output terminal either to $V_{cc}$ or ground, depending on its state. For a gate outputting a "high" logic level, the combination of $Q_3$ saturated and $Q_4$ cutoff is analogous to a double-throw switch in the "$V_{cc}$" position, providing a path for current through a grounded load:

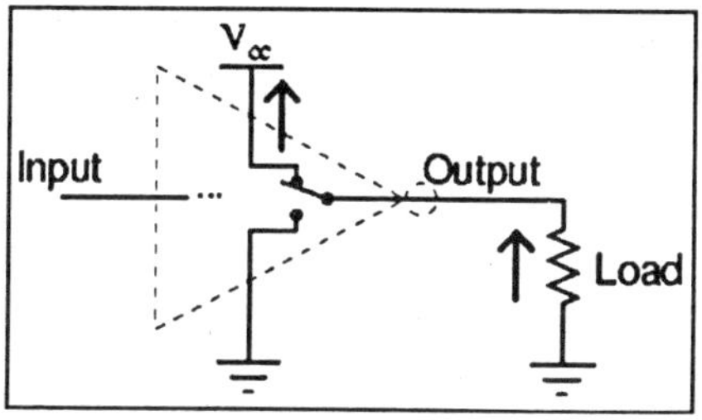


**Fig.** Simplified Gate Circuit SourcingCurrent.

Please note that this two-position switch shown inside the gate symbol is representative of transistors $Q_3$ and $Q_4$ alternately connecting the output terminal to $V_{cc}$ or ground, not of the switch previously shown sending an input signal to the gate!

Conversely, when a gate circuit is outputting a "low" logic level to a load, it is analogous to the double-throw switch being set in the "ground" position.

Current will then be going the other way if the load resistance connects to $V_{cc}$: from ground, through the emitter of $Q_4$, out the output terminal, through the load resistance, and back to $V_{cc}$. In this condition, the gate is said to be sinking current:

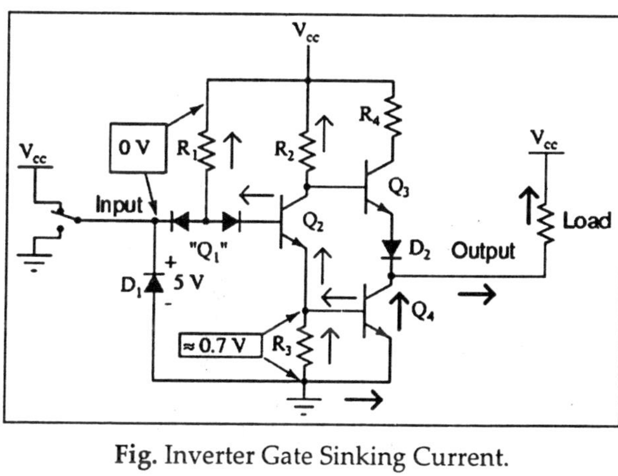


**Fig.** Inverter Gate Sinking Current.

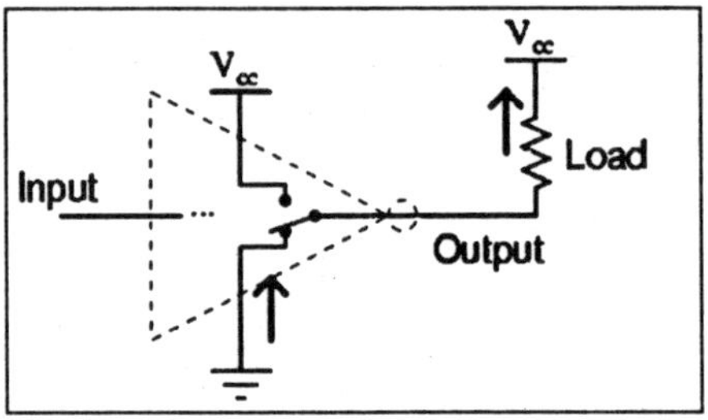


**Fig.** Simplified Gate Circuit Sinking Current.

The combination of $Q_3$ and $Q_4$ working as a "push-pull" transistor pair (otherwise known as a totem pole output) has the ability to either source current (draw in current to $V_{cc}$) or sink current (output current from ground) to a load. However, a standard TTL gate input never needs current to be sourced, only sunk. That is, since a TTL gate input naturally assumes a high state if left floating, any gate output driving a TTL input need only sink current to provide a "0" or "low" input, and need not source current to provide a "1" or a "high" logic level at the input of the receiving gate:

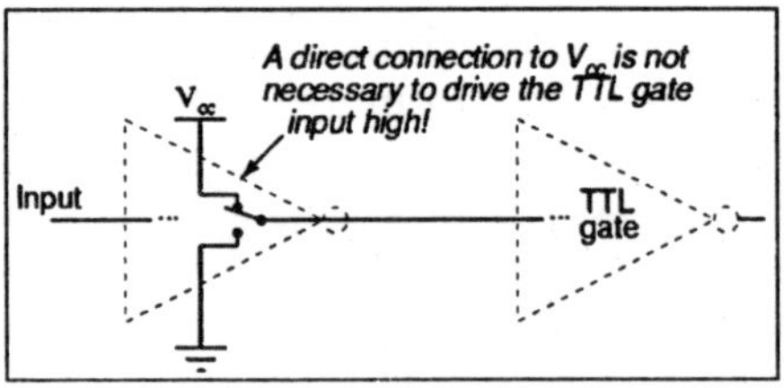


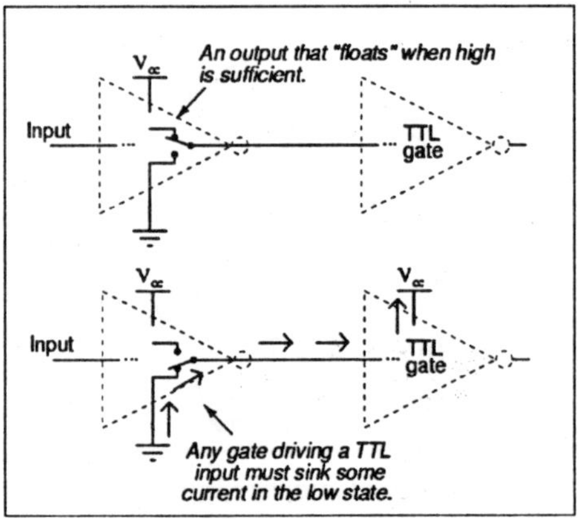

This means we have the option of simplifying the output stage of a gate circuit so as to eliminate $Q_3$ altogether. The result is known as an open-collector output:

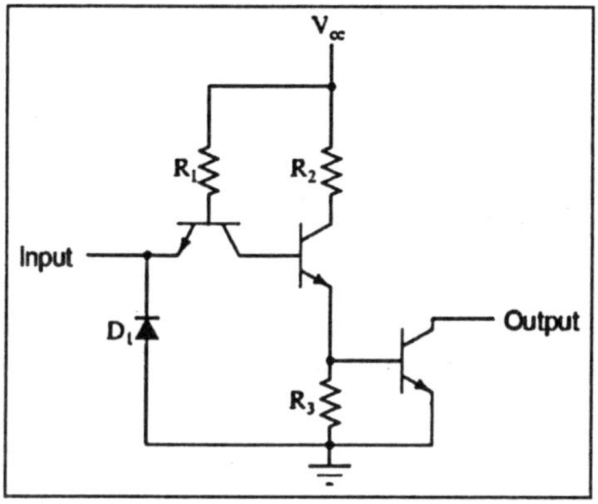


**Fig.** Inverter Circuit with Open Collector Output.

To designate open-collector output circuitry within a standard gate symbol, a special marker is used. Shown here is the symbol for an inverter gate with open-collector output:

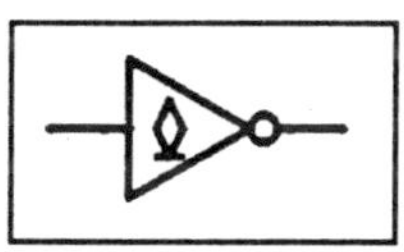

**Fig.** Inverter with Open Collector Output.

Please keep in mind that the "high" default condition of a floating gate input is only true for TTL circuitry, and not necessarily for other types, especially for logic gates constructed of field-effect transistors.

*Review*:

- An inverter, or NOT, gate is one that outputs the opposite state as what is input. That is, a "low" input (0) gives a "high" output (1), and visa-versa.
- Gate circuits constructed of resistors and bipolar transistors as illustrated in this section are called TTL. TTL is an acronym standing for Transistor-to-Transistor Logic. There are other design methodologies used in gate circuits, some which use field-effect transistors rather than bipolar transistors.

- A gate is said to be sourcing current when it provides a path for current between the output terminal and the positive side of the DC power supply ($V_{cc}$). In other words, it is connecting the output terminal to the power source (+V).
- A gate is said to be sinking current when it provides a path for current between the output terminal and ground. In other words, it is grounding (sinking) the output terminal.
- Gate circuits with totem pole output stages are able to both source and sink current. Gate circuits with open-collector output stages are only able to sink current, and not source current. Open-collector gates are practical when used to drive TTL gate inputs because TTL inputs don't require current sourcing.

## THE "BUFFER" GATE

If we were to connect two inverter gates together so that the output of one fed into the input of another, the two inversion functions would "cancel" each other out so that there would be no inversion from input to final output:

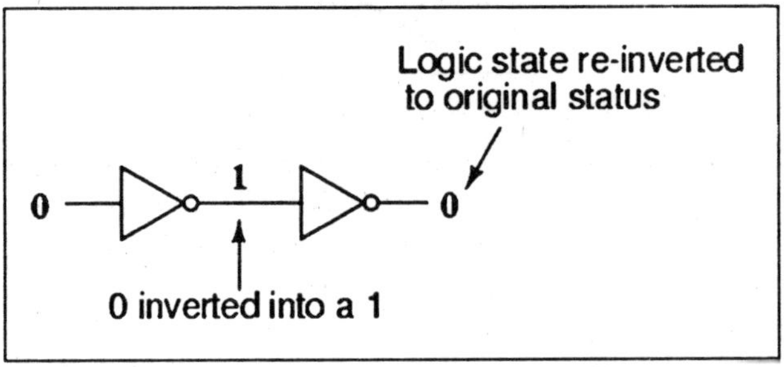


While this may seem like a pointless thing to do, it does have practical application. Remember that gate circuits are signal amplifiers, regardless of what logic function they may perform.

A weak signal source (one that is not capable of sourcing or sinking very much current to a load) may be boosted by means of two inverters like the pair shown in the previous illustration.

The logic level is unchanged, but the full current-sourcing or -sinking capabilities of the final inverter are available to drive a load resistance if needed.

For this purpose, a special logic gate called a buffer is manufactured to perform the same function as two inverters. Its symbol is simply a triangle, with no inverting "bubble" on the output terminal:

*"Buffer" gate*

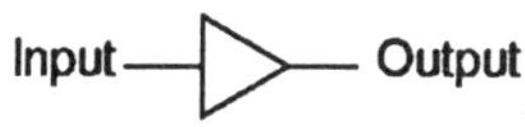


| Input | Output |
|---|---|
| 0 | 0 |
| 1 | 1 |

The internal schematic diagram for a typical open-collector buffer is not much different from that of a simple inverter: only one more common-emitter transistor stage is added to re-invert the output signal.

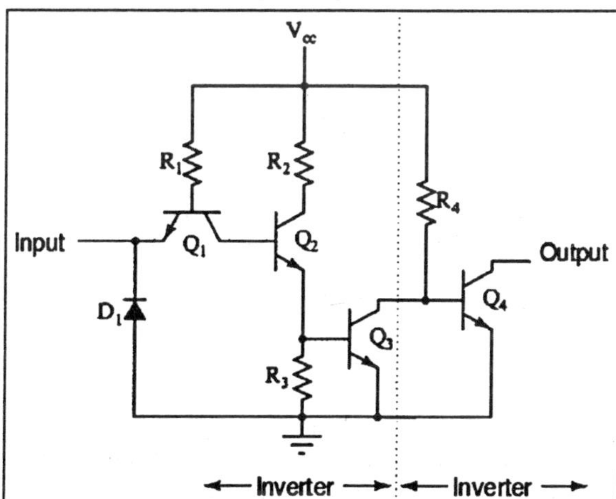


**Fig.** Buffer Circuit with Open-Collector Output.

Let's analyse this circuit for two conditions: an input logic level of "1" and an input logic level of "0." First, a "high" (1) input:

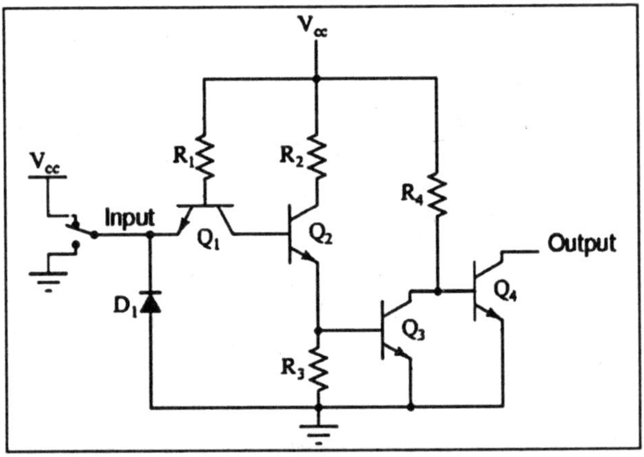


As before with the inverter circuit, the "high" input causes no conduction through the left steering diode of $Q_1$ (emitter-to-base PN junction). All of $R_1$'s current goes through the base of transistor $Q_2$, saturating it:

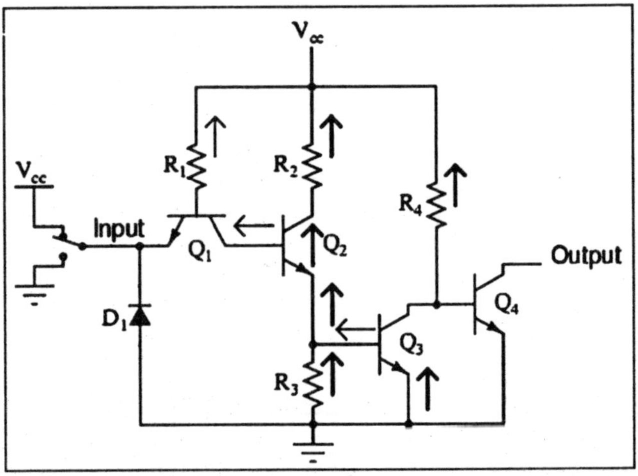


Having $Q_2$ saturated causes $Q_3$ to be saturated as well, resulting in very little voltage dropped between the base and emitter of the final output transistor $Q_4$.

Thus, $Q_4$ will be in cutoff mode, conducting no current. The output terminal will be floating (neither connected to ground nor $V_{cc}$), and this will be equivalent to a "high" state on the input of the next TTL gate that this one feeds in to. Thus, a "high" input gives a "high" output. With a "low" input signal (input terminal grounded), the analysis looks something like this:

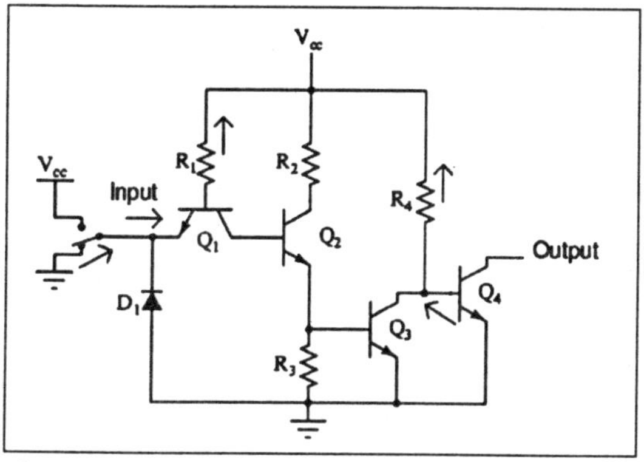


All of $R_1$'s current is now diverted through the input switch, thus eliminating base current through $Q_2$. This forces transistor $Q_2$ into cutoff so that no base current goes through $Q_3$ either. With $Q_3$ cutoff as well, $Q_4$ is will be saturated by the current through resistor $R_4$, thus connecting the output terminal to ground, making it a "low" logic level. Thus, a "low" input gives a "low" output. The schematic diagram for a buffer circuit with totem pole output transistors is a bit more complex, but the basic principles, and certainly the truth table, are the same as for the open-collector circuit:

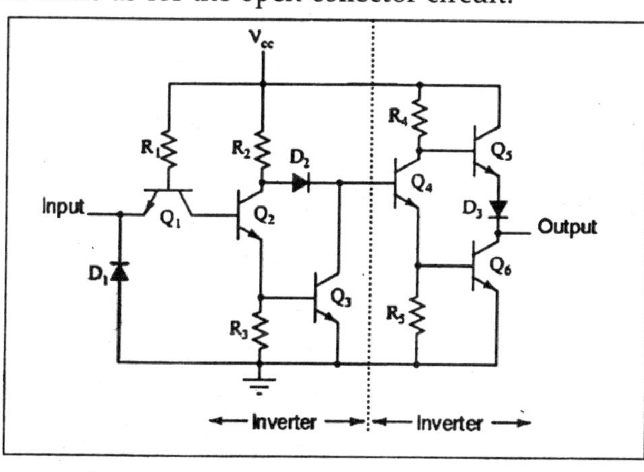


**Fig.** Buffer Circuit with Totem Pole Output.

*Review*:

- Two inverter, or NOT, gates connected in "series" so as to invert, then re-invert, a binary bit perform the function of a buffer. Buffer gates merely serve the purpose of signal amplification: taking a "weak" signal source that isn't capable of sourcing or sinking much current, and boosting the current capacity of the signal so as to be able to drive a load.
- Buffer circuits are symbolized by a triangle symbol with no inverter "bubble."
- Buffers, like inverters, may be made in open-collector output or totem pole output forms.

## MULTIPLE-INPUT GATES

Inverters and buffers exhaust the possibilities for single-input gate circuits. What more can be done with a single logic signal but to buffer it or invert it? To explore more logic gate possibilities, we must add more input terminals to the circuits.

Adding more input terminals to a logic gate increases the number of input state possibilities. With a single-input gate such as the inverter or buffer, there can only be two possible input states: either the input is "high" (1) or it is "low" (0).

As was mentioned previously in this chapter, a two input gate has four possibilities (00, 01, 10, and 11).

A three-input gate has eight possibilities (000, 001, 010, 011, 100, 101, 110, and 111) for input states. The number of possible input states is equal to two to the power of the number of inputs:

Number of possible input states = $2^n$

Where,

$$n = \text{Number of inputs}$$

This increase in the number of possible input states obviously allows for more complex gate behaviour. Now, instead of merely inverting or amplifying (buffering) a single "high" or "low" logic level, the output of the gate will be determined by whatever combination of 1's and 0's is present at the input terminals.

Since so many combinations are possible with just a few input terminals, there are many different types of multiple-input gates, unlike single-input gates which can only be inverters or buffers. Each basic gate type will be presented in this section, showing its standard symbol, truth table, and practical operation.

## THE AND GATE

One of the easiest multiple-input gates to understand is the AND gate, so-called because the output of this gate will be "high" (1) if and only if all inputs (first input and the second input and...) are "high" (1). If any input(s) are "low" (0), the output is guaranteed to be in a "low" state as well.

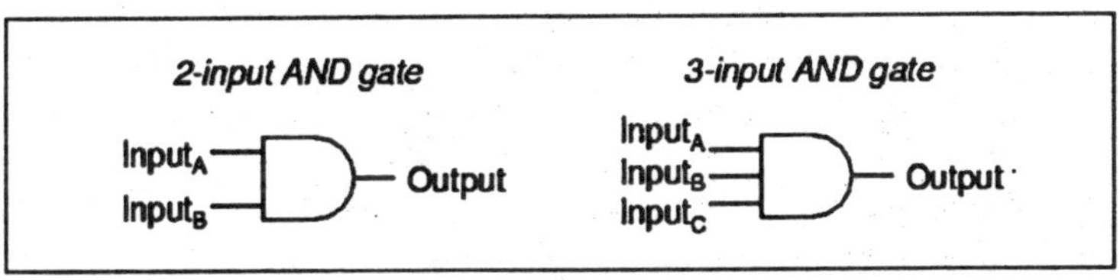


In case you might have been wondering, AND gates are made with more than three inputs, but this is less common than the simple two-input variety.

*A two-input AND gate's truth table looks like this:*

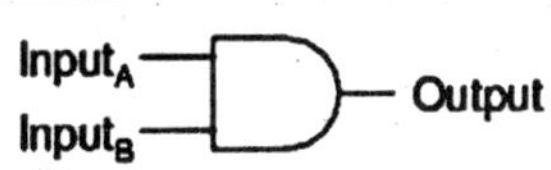


| A | B | Output |
|---|---|---|
| 0 | 0 | 0 |
| 0 | 1 | 0 |
| 1 | 0 | 0 |
| 1 | 1 | 1 |

What this truth table means in practical terms is shown in the following sequence of illustrations, with the 2-input AND gate subjected to all possibilities of input logic levels. An LED (Light-Emitting Diode) provides visual indication of the output logic level:

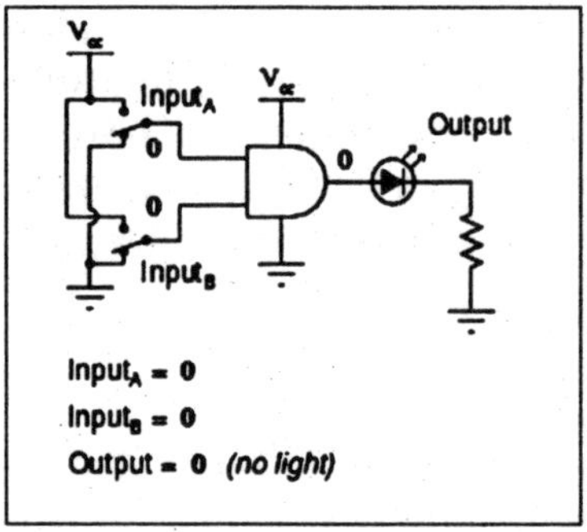


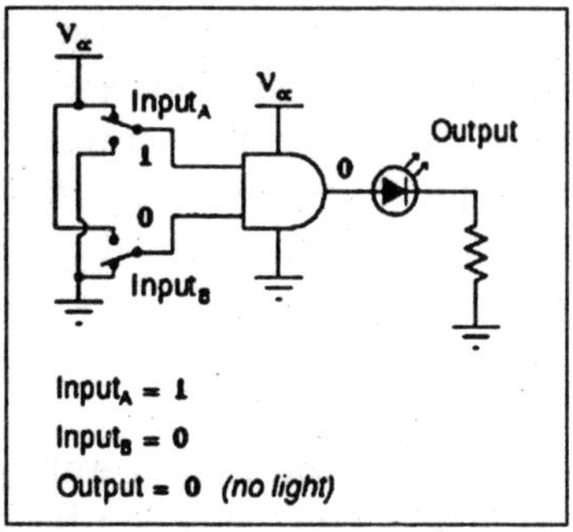


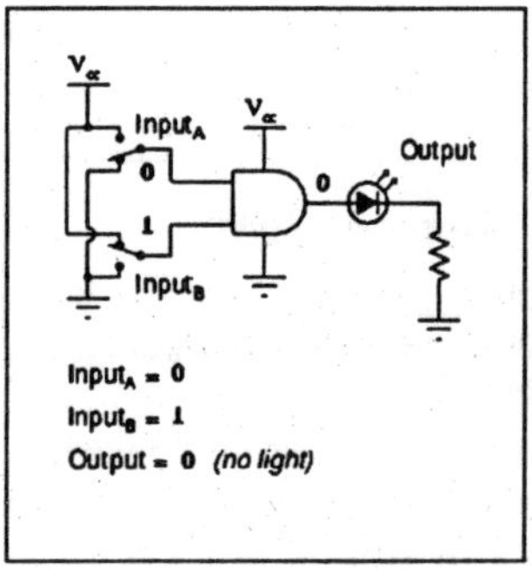


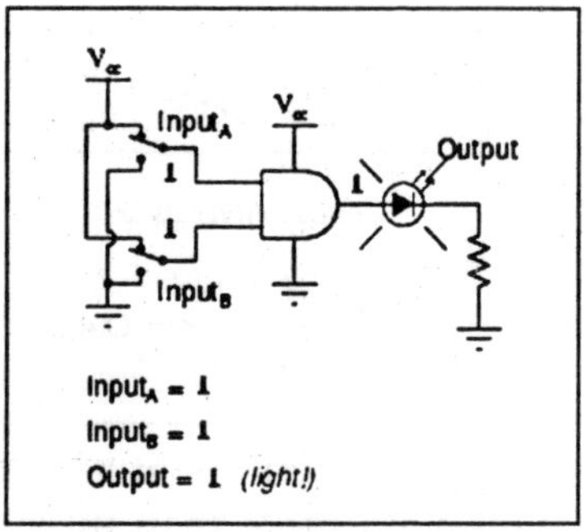


It is only with all inputs raised to "high" logic levels that the AND gate's output goes "high," thus energizing the LED for only one out of the four input combination states.

## THE NAND GATE

A variation on the idea of the AND gate is called the NAND gate. The word "NAND" is a verbal contraction of the words NOT and AND. Essentially, a NAND gate behaves the same as an AND gate with a NOT (inverter) gate connected to the output terminal. To symbolize this output signal inversion, the NAND gate symbol has a bubble on the output line. The truth table for a NAND gate is as one might expect, exactly opposite as that of an AND gate:

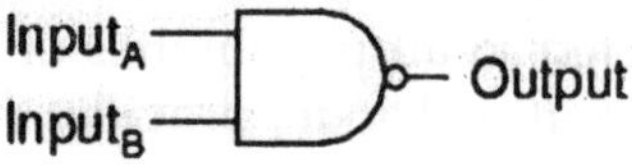

| A | B | Output |
|---|---|---|
| 0 | 0 | 1 |
| 0 | 1 | 1 |
| 1 | 0 | 1 |
| 1 | 1 | 0 |

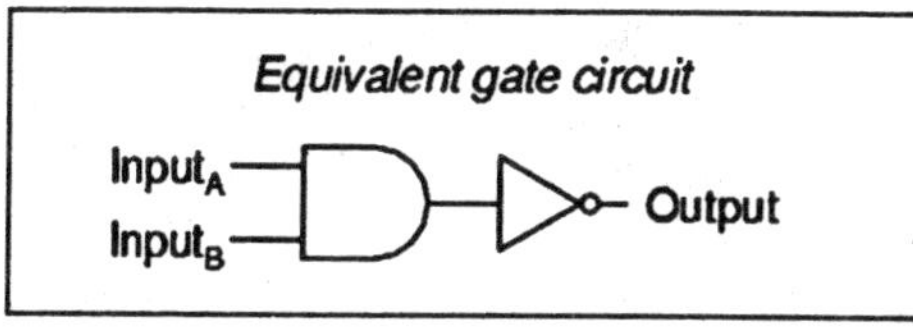


As with AND gates, NAND gates are madė with more than two inputs. In such cases, the same general principle applies: the output will be "low" (0) if and only if all inputs are "high" (1). If any input is "low" (0), the output will go "high" (1).

## THE OR GATE

Our next gate to investigate is the OR gate, so-called because the output of this gate will be "high" (1) if any of the inputs (first input or the second input or...) are "high" (1). The output of an OR gate goes "low" (0) if and only if all inputs are "low" (0).

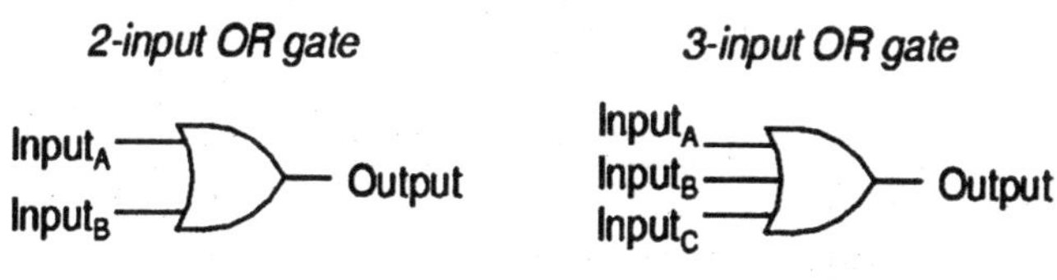


A two-input OR gate's truth table looks like this:

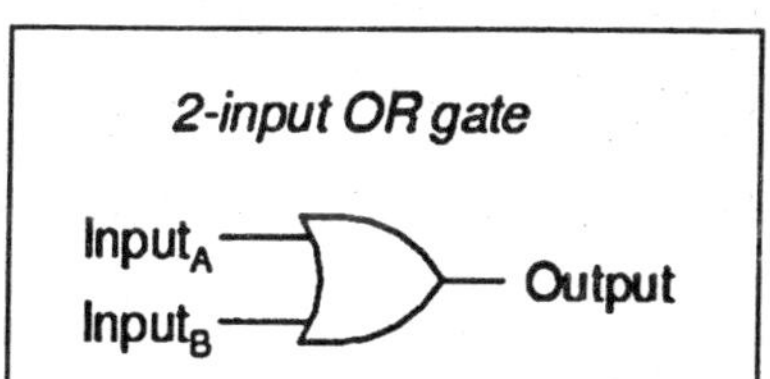


The following sequence of illustrations demonstrates the

OR gate's function, with the 2-inputs experiencing all possible logic levels. An LED (Light-Emitting Diode) provides visual indication of the gate's output logic level:

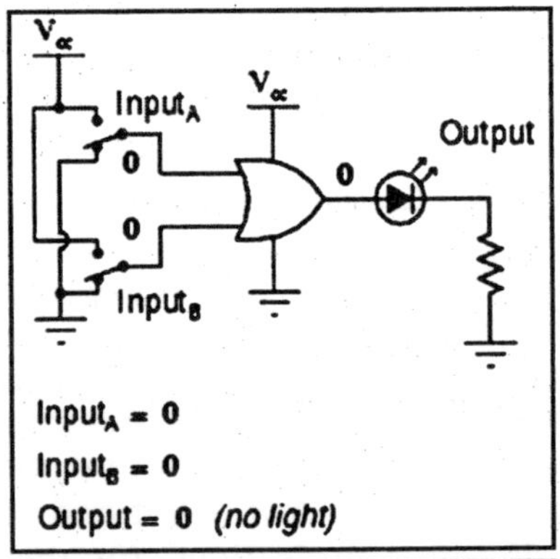


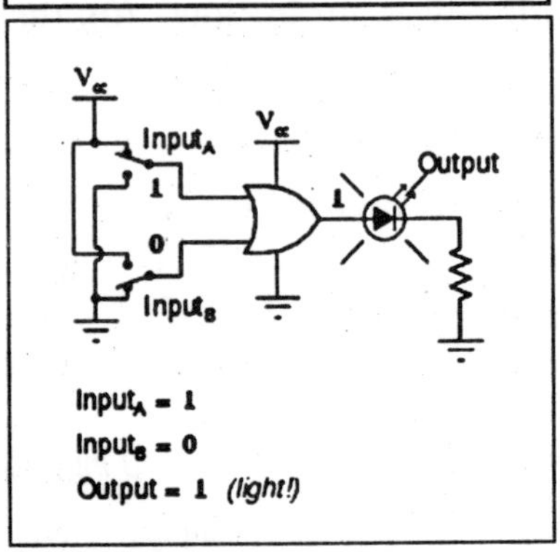


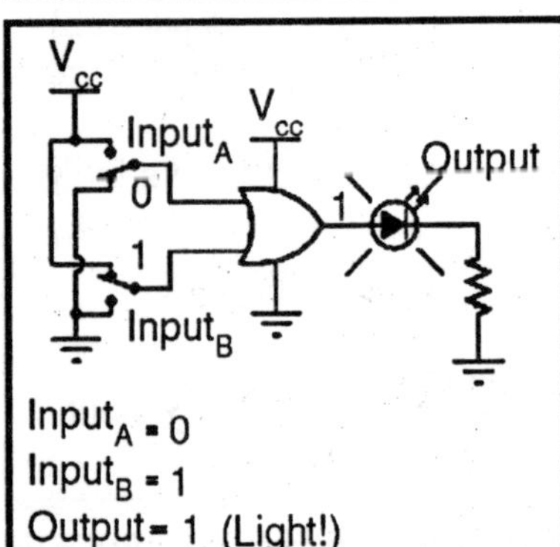

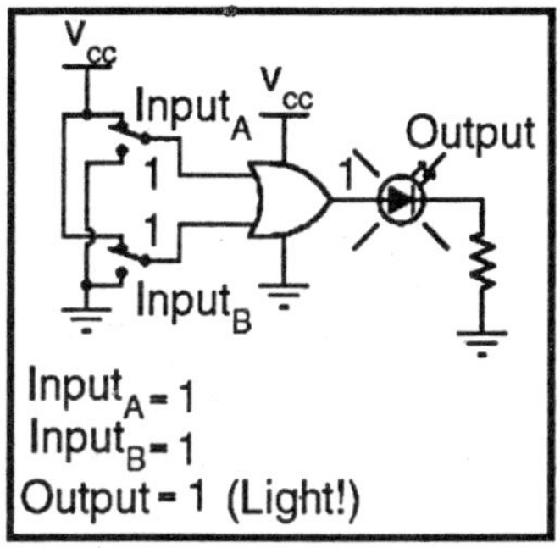


A condition of any input being raised to a "high" logic level makes the OR gate's output go "high," thus energizing the LED for three out of the four input combination states.

## THE NOR GATE

As you might have suspected, the NOR gate is an OR gate with its output inverted, just like a NAND gate is an AND gate with an inverted output.

2-Input NOR Gate

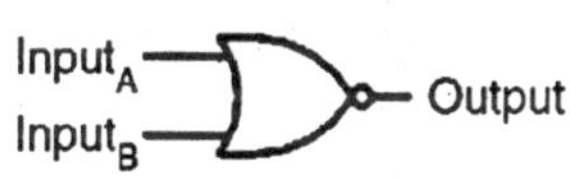


| A | B | Output |
|---|---|---|
| 0 | 0 | 1 |
| 0 | 1 | 0 |
| 1 | 0 | 0 |
| 1 | 1 | 0 |

Equivalent Gate Circuit

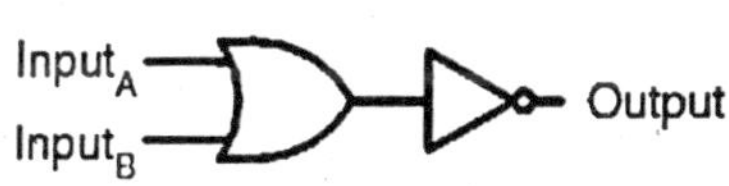


NOR gates, like all the other multiple-input gates seen thus far, can be manufactured with more than two inputs.

*Still, the same logical principle applies*: The output goes "low" (0) if any of the inputs are made "high" (1). The output is "high" (1) only when all inputs are "low" (0).

## THE NEGATIVE-AND GATE

A Negative-AND gate functions the same as an AND gate with all its inputs inverted (connected through NOT gates). In keeping with standard gate symbol convention, these inverted inputs are signified by bubbles.

Contrary to most peoples' first instinct, the logical behaviour of a Negative-AND gate is not the same as a NAND gate. Its truth table, actually, is identical to a NOR gate:

2-Input Negative-AND Gate

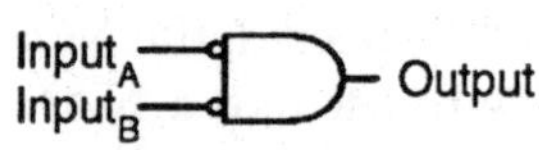


| A | B | Output |
|---|---|---|
| 0 | 0 | 1 |
| 0 | 1 | 0 |
| 1 | 0 | 0 |
| 1 | 1 | 0 |

Equivalent Gate Circuits

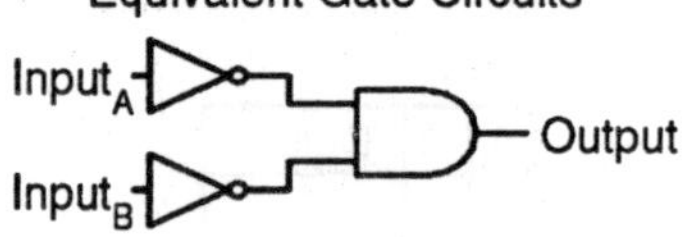

The Negative-OR Gate [H]

## THE NEGATIVE-OR GATE

Following the same pattern, a Negative-OR gate functions the same as an OR gate with all its inputs inverted. In keeping with standard gate symbol convention, these inverted inputs are signified by bubbles.

The behaviour and truth table of a Negative-OR gate is the same as for a NAND gate:

2-Input Negative-OR Gate

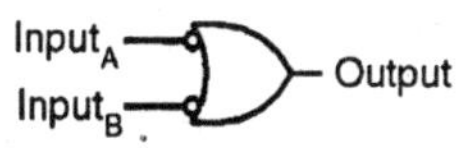


| A | B | Output |
|---|---|---|
| 0 | 0 | 1 |
| 0 | 1 | 1 |
| 1 | 0 | 1 |
| 1 | 1 | 0 |

Equivalent Gate Circuits

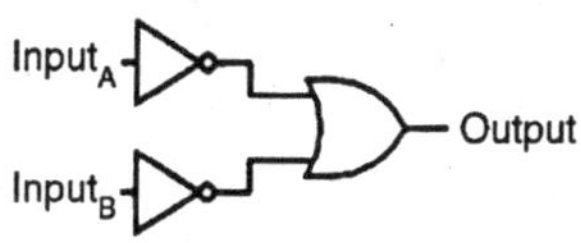


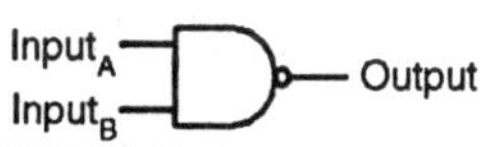


## THE EXCLUSIVE-OR GATE

The last six gate types are all fairly direct variations on three basic functions: AND, OR, and NOT. The Exclusive-OR gate, however, is something quite different.

Exclusive-OR gates output a "high" (1) logic level if the inputs are at different logic levels, either 0 and 1 or 1 and 0. Conversely, they output a "low" (0) logic level if the inputs are at the same logic levels. The Exclusive-OR (sometimes called XOR) gate has both a symbol and a truth table pattern that is unique:

Exclusive-OR Gate

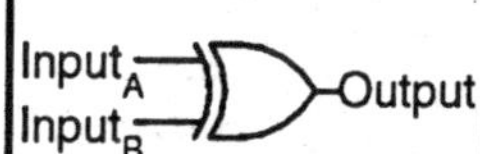


| A | B | Output |
|---|---|---|
| 0 | 0 | 0 |
| 0 | 1 | 1 |
| 1 | 0 | 1 |
| 1 | 1 | 0 |

There are equivalent circuits for an Exclusive-OR gate made up of AND, OR, and NOT gates, just as there were for NAND, NOR, and the negative-input gates.

A rather direct approach to simulating an Exclusive-OR gate is to start with a regular OR gate, then add additional gates to inhibit the output from going "high" (1) when both inputs are "high" (1):

Exclusive-OR Equivalent Circuit

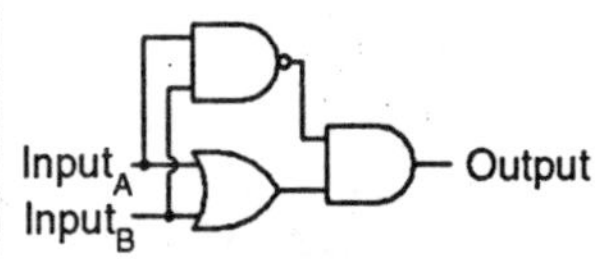


| A | B | Output |
|---|---|---|
| 0 | 0 | 0 |
| 0 | 1 | 1 |
| 1 | 0 | 1 |
| 1 | 1 | 0 |

In this circuit, the final AND gate acts as a buffer for the output of the OR gate whenever the NAND gate's output is high, which it is for the first three input state combinations (00, 01, and 10). However, when both inputs are "high" (1), the NAND gate outputs a "low" (0) logic level, which forces the final AND gate to produce a "low" (0) output.

Another equivalent circuit for the Exclusive-OR gate uses a strategy of two AND gates with inverters, set up to generate "high" (1) outputs for input conditions 01 and 10. A final OR gate then allows either of the AND gates' "high" outputs to create a final "high" output:

Exclusive-OR Equivalent Circuit

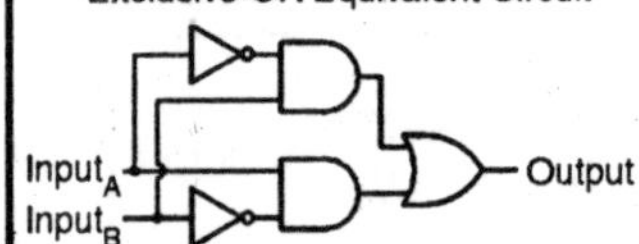


| A | B | Output |
|---|---|---|
| 0 | 0 | 0 |
| 0 | 1 | 1 |
| 1 | 0 | 1 |
| 1 | 1 | 0 |

Exclusive-OR gates are very useful for circuits where two

or more binary numbers are to be compared bit-for-bit, and also for error detection (parity check) and code conversion (binary to Grey and visa-versa).

## THE EXCLUSIVE-NOR GATE

Finally, our last gate for analysis is the Exclusive-NOR gate, otherwise known as the XNOR gate. It is equivalent to an Exclusive-OR gate with an inverted output. The truth table for this gate is exactly opposite as for the Exclusive-OR gate:

Excluive-NOR Gate

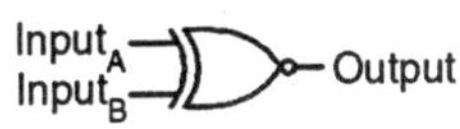


| A | B | Output |
|---|---|---|
| 0 | 0 | 1 |
| 0 | 1 | 0 |
| 1 | 0 | 0 |
| 1 | 1 | 1 |

Equivalent Gate Circuit

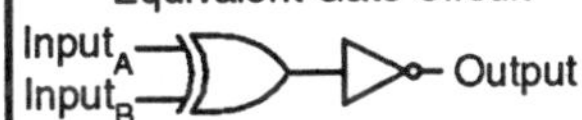


As indicated by the truth table, the purpose of an Exclusive-NOR gate is to output a "high" (1) logic level whenever both inputs are at the same logic levels (either 00 or 11).

*Review*:

- Rule for an AND gate: output is "high" only if first input and second input are both "high."
- Rule for an OR gate: output is "high" if input A or input B are "high."
- Rule for a NAND gate: output is not "high" if both the first input and the second input are "high."
- Rule for a NOR gate: output is not "high" if either the first input or the second input are "high."
- A Negative-AND gate behaves like a NOR gate.
- A Negative-OR gate behaves like a NAND gate.
- Rule for an Exclusive-OR gate: output is "high" if the input logic levels are different.

- Rule for an Exclusive-NOR gate: output is "high" if the input logic levels are the same.

## TTL NAND AND AND GATES

Suppose we altered our basic open-collector inverter circuit, adding a second input terminal just like the first:

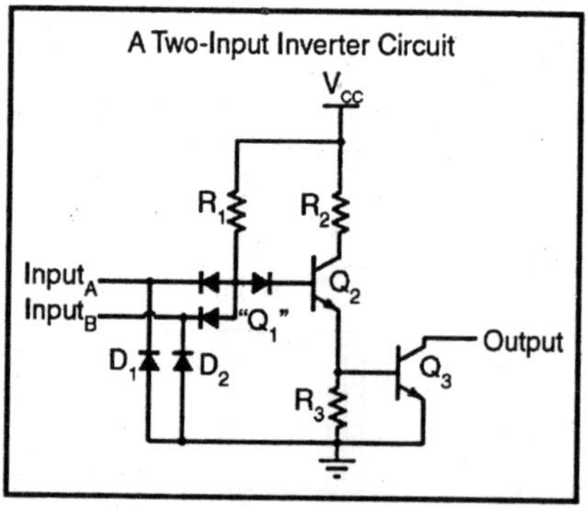


This schematic illustrates a real circuit, but it isn't called a "two-input inverter." Through analysis we will discover what this circuit's logic function is and correspondingly what it should be designated as.

Just as in the case of the inverter and buffer, the "steering" diode cluster marked "$Q_1$" is actually formed like a transistor, even though it isn't used in any amplifying capacity. Unfortunately, a simple NPN transistor structure is inadequate to simulate the three PN junctions necessary in this diode network, so a different transistor (and symbol) is needed. This transistor has one collector, one base, and two emitters, and in the circuit it looks like this:

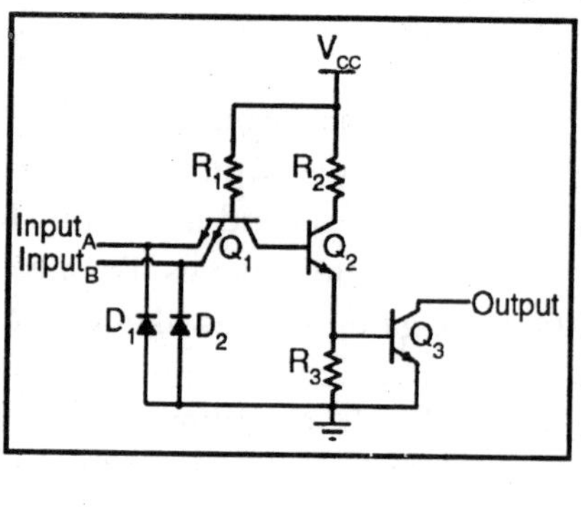

In the single-input (inverter) circuit, grounding the input resulted in an output that assumed the "high" (1) state. In the case of the open-collector output configuration, this "high" state was simply "floating." Allowing the input to float (or be connected to $V_{cc}$) resulted in the output becoming grounded, which is the "low" or 0 state. Thus, a 1 in resulted in a 0 out, and visa-versa. Since this circuit bears so much resemblance to the simple inverter circuit, the only difference being a second input terminal connected in the same way to the base of transistor $Q_2$, we can say that each of the inputs will have the same effect on the output. Namely, if either of the inputs are grounded, transistor $Q_2$ will be forced into a condition of cutoff, thus turning $Q_3$ off and floating the output (output goes "high"). The following series of illustrations shows this for three input states (00, 01, and 10):

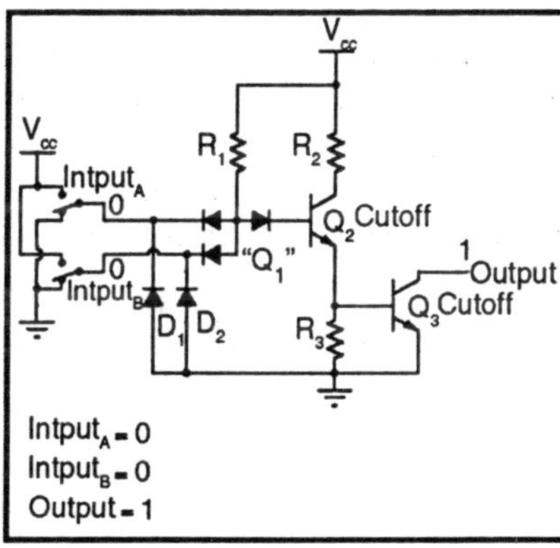


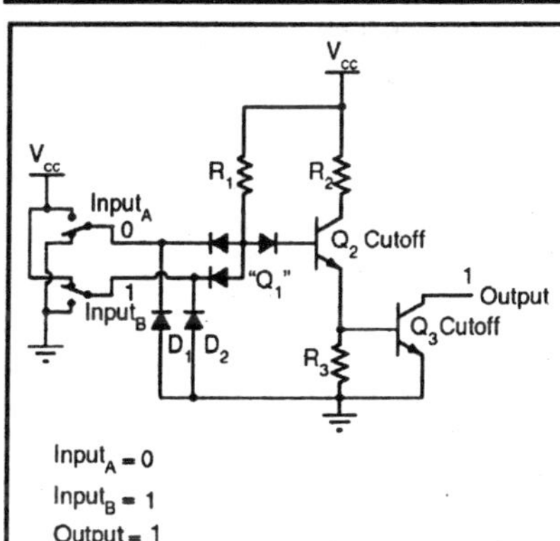

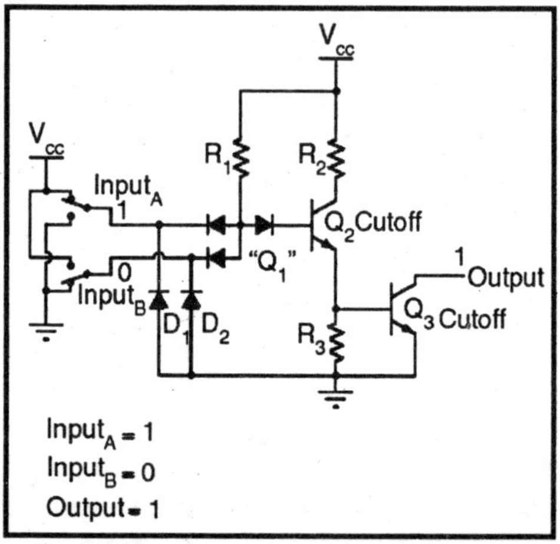


In any case where there is a grounded ("low") input, the output is guaranteed to be floating ("high"). Conversely, the only time the output will ever go "low" is if transistor $Q_3$ turns on, which means transistor $Q_2$ must be turned on (saturated), which means neither input can be diverting $R_1$ current away from the base of $Q_2$.

The only condition that will satisfy this requirement is when both inputs are "high" (1):

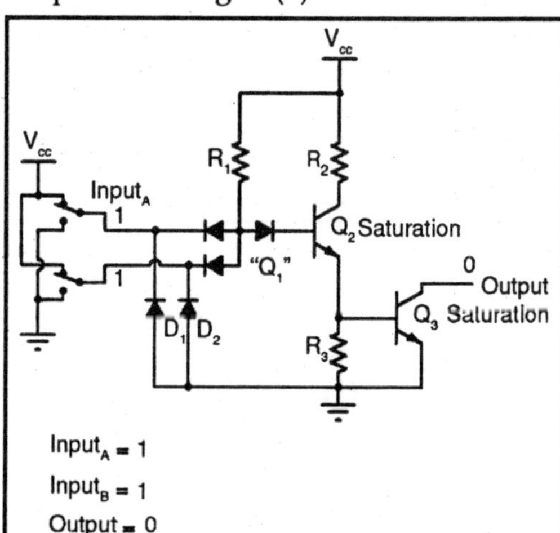


Collecting and tabulating these results into a truth table, we see that the pattern matches that of the NAND gate:

NAND Gate

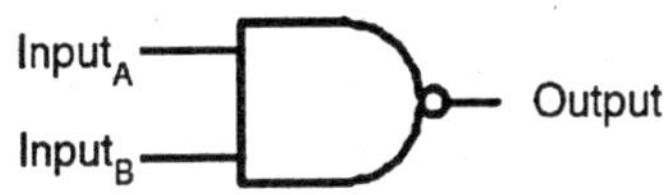


| A | B | Output |
|---|---|---|
| 0 | 0 | 1 |
| 0 | 1 | 1 |
| 1 | 0 | 1 |
| 1 | 1 | 0 |

In the earlier section on NAND gates, this type of gate was created by taking an AND gate and increasing its complexity by adding an inverter (NOT gate) to the output. However, when we examine this circuit, we see that the NAND function is actually the simplest, most natural mode of operation for this TTL design. To create an AND function using TTL circuitry, we need to increase the complexity of this circuit by adding an inverter stage to the output, just like we had to add an additional transistor stage to the TTL inverter circuit to turn it into a buffer:

AND Gate With Open-Collector Output

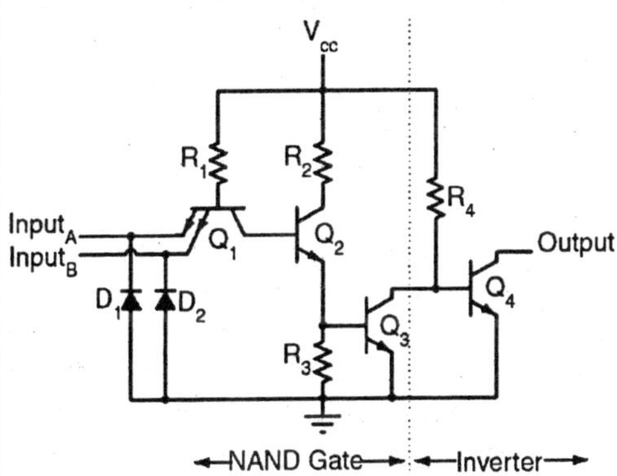

The truth table and equivalent gate circuit (an inverted-output NAND gate) are shown here:

AND Gate

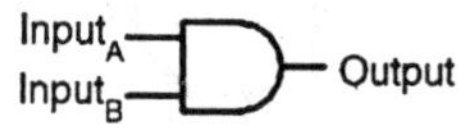


| A | B | Output |
|---|---|---|
| 0 | 0 | 0 |
| 0 | 1 | 0 |
| 1 | 0 | 0 |
| 1 | 1 | 1 |

Equivalent Circuit

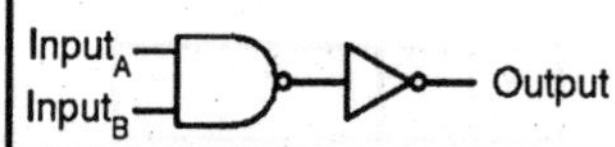


Of course, both NAND and AND gate circuits may be designed with totem-pole output stages rather than open-collector.

*Review*:

- A TTL NAND gate can be made by taking a TTL inverter circuit and adding another input.
- An AND gate may be created by adding an inverter stage to the output of the NAND gate circuit.

## TTL NOR AND OR GATES

Let's examine the following TTL circuit and analyse its operation:

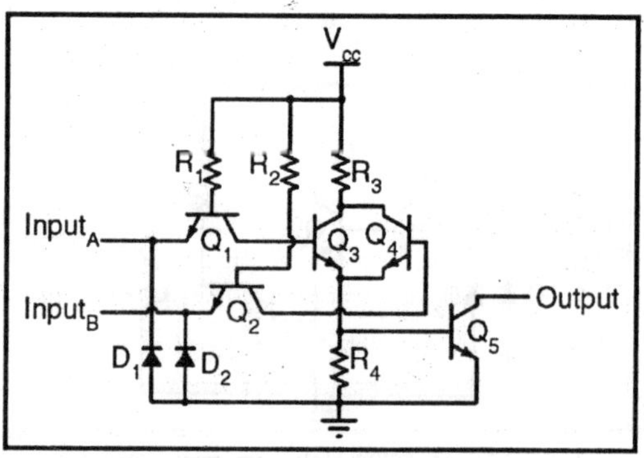


Transistors $Q_1$ and $Q_2$ are both arranged in the same

manner that we've seen for transistor $Q_1$ in all the other TTL circuits. Rather than functioning as amplifiers, $Q_1$ and $Q_2$ are both being used as two-diode "steering" networks. We may replace $Q_1$ and $Q_2$ with diode sets to help illustrate:

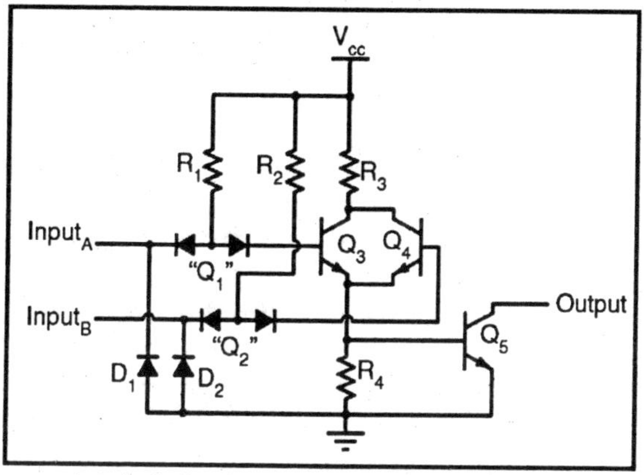


If input A is left floating (or connected to $V_{cc}$), current will go through the base of transistor $Q_3$, saturating it. If input A is grounded, that current is diverted away from $Q_3$'s base through the left steering diode of "$Q_1$," thus forcing $Q_3$ into cutoff.

The same can be said for input B and transistor $Q_4$: the logic level of input B determines $Q_4$'s conduction: either saturated or cutoff.

Notice how transistors $Q_3$ and $Q_4$ are paralleled at their collector and emitter terminals.

In essence, these two transistors are acting as paralleled switches, allowing current through resistors $R_3$ and $R_4$ according to the logic levels of inputs A and B.

If any input is at a "high" (1) level, then at least one of the two transistors ($Q_3$ and/or $Q_4$) will be saturated, allowing current through resistors $R_3$ and $R_4$, and turning on the final output transistor $Q_5$ for a "low" (0) logic level output.

The only way the output of this circuit can ever assume a "high" (1) state is if both $Q_3$ and $Q_4$ are cutoff, which means both inputs would have to be grounded, or "low" (0).

This circuit's truth table, then, is equivalent to that of the NOR gate:

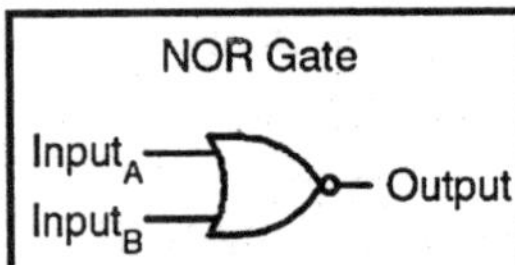


| A | B | Output |
|---|---|---|
| 0 | 0 | 1 |
| 0 | 1 | 0 |
| 1 | 0 | 0 |
| 1 | 1 | 0 |

In order to turn this NOR gate circuit into an OR gate, we would have to invert the output logic level with another transistor stage, just like we did with the NAND-to-AND gate example:

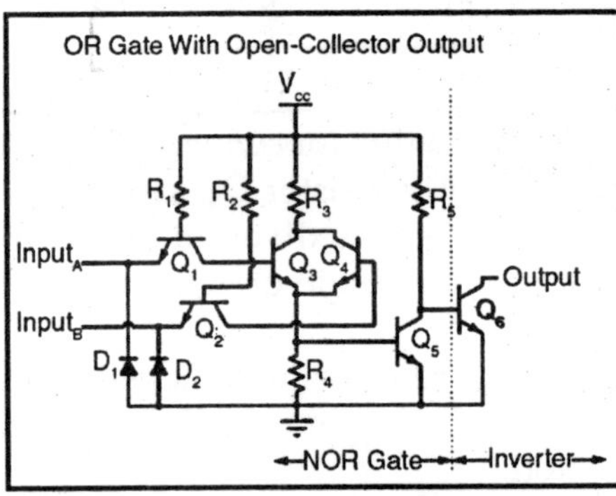


The truth table and equivalent gate circuit (an inverted-output NOR gate) are shown here:

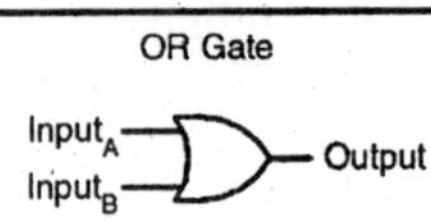


| A | B | Output |
|---|---|---|
| 0 | 0 | 0 |
| 0 | 1 | 1 |
| 1 | 0 | 1 |
| 1 | 1 | 1 |

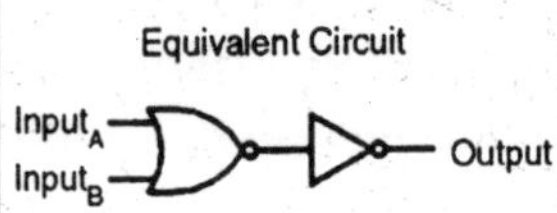

cache memory, for which it is perfectly suited; cache memory needs to be very fast, and not very large.

SRAM is manufactured in a way rather similar to how processors are: highly-integrated transistor patterns photo-etched into silicon. Each SRAM bit is comprised of between four and six transistors, which is why SRAM takes up much more space compared to DRAM, which uses only one (plus a capacitor). Because an SRAM chip is comprised of thousands or millions of identical cells, it is much easier to make than a CPU, which is a large die with a non-repetitive structure. This is one reason why RAM chips cost much less than processors do. See this discussion of how processors are manufactured; this process is similar (but simplified somewhat) for making memory circuits.

## Dynamic RAM (DRAM)

Dynamic RAM is a type of RAM that only holds its data if it is continuously accessed by special logic called a refresh circuit. Many hundreds of times each second, this circuitry reads the contents of each memory cell, whether the memory cell is being used at that time by the computer or not. Due to the way in which the cells are constructed, the reading action itself refreshes the contents of the memory. If this is not done regularly, then the DRAM will lose its contents, even if it continues to have power supplied to it. This refreshing action is why the memory is called dynamic.

All PCs use DRAM for their main system memory, instead of SRAM, even though DRAMs are slower than SRAMs and require the overhead of the refresh circuitry. It may seem weird to want to make the computer's memory out of something that can only hold a value for a fraction of a second. In fact, DRAMs are both more complicated and slower than SRAMs.

The reason that DRAMs are used is simple: they are much cheaper and take up much less space, typically 1/4 the silicon area of SRAMs or less. To build a 64 MB core memory from SRAMs would be very expensive. The overhead of the refresh circuit is tolerated in order to allow the use of large amounts of inexpensive, compact memory. The refresh circuitry itself is almost never a problem; many years of using DRAM has caused the design of these circuits to be all but perfected.

DRAMs are smaller and less expensive than SRAMs because SRAMs are made from four to six transistors (or more) per bit, DRAMs use only one, plus a capacitor. The capacitor, when energized, holds an electrical charge if the bit contains a "1" or no charge if it contains a "0". The transistor is used to read the contents of the capacitor. The problem with capacitors is that they only hold a charge for a short period of time, and then it fades away. These capacitors are tiny, so their charges fade particularly quickly. This is why the refresh circuitry is needed: to read the contents of every cell and refresh them with a fresh "charge" before the contents fade away and are lost. Refreshing is done by reading every "row" in the memory chip one row at a time; the process of reading the contents of each capacitor re-establishes the charge. For an explanation of how these "rows" are read, and thus how refresh is accomplished, refer to this section describing memory access.

DRAM is manufactured using a similar process to how processors are: a silicon substrate is etched with the patterns that make the transistors and capacitors (and support structures) that comprise each bit. DRAM costs much less than a processor because it is a series of simple, repeated structures, so there isn't the complexity of making a single chip with several million individually-located transistors. See here for details on how processors are manufactured; the principles for DRAM manufacture are similar.

There are many different kinds of specific DRAM technologies and speeds that they are available in. These have evolved over many years of using DRAM for system memory, and are discussed in more detail in other sections.

Next: Memory Speed, Access and Timing

## Read-only memory (ROM)

There is a type of memory that stores data without electrical current; it is the ROM (Read Only Memory) or is sometimes called non-volatile memory as it is not erased when the system is switched off.

This type of memory lets you stored the data needed to start up the computer. Indeed, this information cannot be stored on the hard disk since the disk parameters (vital for its initialisation) are part of these data which are essential for booting. Different ROM-type memories contain these essential start-up data, i.e:

- The BIOS is a programme for controlling the system's main input-output interfaces, hence the name BIOS ROM which is sometimes given to the read-only memory chip of the mother board which hosts it.
- The bootstrap loader: a programme for loading (random access) memory into the operating system and launching it. This generally seeks the operating system on the floppy drive then on the hard disk, which allows the operating system to be launched from a system floppy disk in the event of malfunction of the system installed on the hard disk.
- The CMOS Setup is the screen displayed when the computer starts up and which is used to amend the system parameters (often wrongly referred to as BIOS).
- The Power-On Self Test (POST), a programme that runs automatically when the system is booted, thus allowing the system to be tested (this is why the system "counts" the RAM at start-up).

Given that ROM are much slower than RAM memories (access time for a ROM is around 150 ns whereas for SDRAM it is around 10 ns), the instructions given in the ROM are sometimes copied to the RAM at start-up; this is known as shadowing, though is usually referred to as shadow memory).

### *Types of ROM*

ROM memories have gradually evolved from fixed read-only memories to memories than can be programmed and then re-programmed.

#### *ROM*

The first ROMs were made using a procedure that directly

writes the binary data in a silicon plate using a mask. This procedure is now obsolete.

*PROM*

PROM (Programmable Read Only Memory) memories were developed at the end of the 70s by a company called Texas Instruments. These memories are chips comprising thousands of fuses (or diodes) that can be "burnt" using a device called a " ROM programmer", applying high voltage (12V) to the memory boxes to be marked. The fuses thus burnt correspond to 0 and the others to 1.

*EPROM*

EPROM (Erasable Programmable Read Only Memory) memories are PROMs that can be deleted. These chips have a glass panel that lets ultra-violet rays through. When the chip is subjected to ultra-violet rays with a certain wavelength, the fuses are reconstituted, meaning that all the memory bits return to 1. This is why this type of PROM is called erasable.

*EEPROM*

EEPROM (Electrically Erasable Read Only Memory memories are also erasable PROMs, but unlike EPROMs, they can be erased by a simple electric current, meaning that they can be erased even when they are in position in the computer.

There is a variant of these memories known as flash memories (also Flash ROM or Flash EPROM). Unlike the classic EEPROMs that use 2 to 3 transistors for each bit to be memorised, the EPROM Flash uses only one transistor. Moreover, the EEPROM may be written and read word by word, while the Flash can be erased only in pages (the size of the pages decreases constantly).

Lastly, the Flash memory is denser, meaning that chips containing several hundred mega octets can be produced. EEPROMs are thus used preferably to memorise configuration data and the Flash memory is used for programmable code (IT programmes).

Once ROM operations are completed, the computer is

ready for normal operational use. For all this to take place, ROM uses circuits in the computer that we have already discussed. In some cases, they are circuits specific for ROM operations. They include: Registers and flip-flops Timing Control signals Internal bus READ-ONLY MEMORY TYPES Types of ROMs include the basic ROM that once manufactured cannot be written on again.

Other types, called programmable read-only memories (PROMs), can be written on again and again. Read-Only Memory ROMs are prepared at the factory. They are not meant to be changed by the user or the technician. They are only to be changed when a newer version is authorized and supplied to replace the old one. Programmable Read-Only memory (PROM) A PROM is a programmable ROM.

Once programmed it acts like a ROM. It can be field-programmed by an authorized technician. Each cell is identified by selecting the row and the column just like locating an address in read/write memory. There are two types of PROMs—erasable and nonerasable. Erasable PROMS can be erased and reprogrammed. Nonerasable PROMS cannot be changed once they are programmed. There are a couple of ways to create or erase the ones in the array; electrically or with ultraviolet (UV) light.

Some PROMS are electrically programmed but erased with UV light. Others are erased electrically and programmed with the UV light. Electrical.— An electric charge can be used to either blow fusible links permanently in a cell or used on a special transistor with two gates. With the special transistors, the gate between the memory cell and the column wire is disabled by the electrical charge. UV light.— UV light is used to erase data in a cell by exposing the IC die to the UV light for a few minutes (usually less than 30 minutes).

UV light can also be used to restore ones to a cell by dissipating the electrical charge that disabled the gate. Electrically alterable or erasable prom (EEPROM or EEPROM).— The EEPROM or EEPROM can be programmed (modified) or erased while it is still in the circuit and used like a nonvolatile read/write memory. EAPROMs/EEPROMs use an electric charge to erase the ones. Some types of EAPROMs/ EEPROMs are more versatile; individual cells can be reprogrammed by reversing the voltage used to create a zero.

There are some timing constraints that cause the part to need more time for erasure or programming than is needed to read data from the part. Some EAPROM/EEPROMs have a word or byte erase mode. ULTRAVIOLET-ERASABLE PROM (UV EPROM OR EPROM).— UV EPROMs/EPROMs trap a charge (1) in the cells to represent the data. To release the charge, the cells are exposed to the UV light for 30 minutes or less. UV EPROMs/EPROMs are usually programmed out of circuit.

Memory modules are interchangeable with other modules of the same type and size in the same computer set. Each module provides a fixed number of memory words with a fixed number of bit positions for each word.

MEMORY ARCHITECTURE— Memories are typically organized in square form so that they have an equal number of rows (x) and columns (y). Each intersection of a row and a column comprises a memory word address. Each memory address will contain a memory word. MEMORY OPERATIONS— Memory opera- tions operate on a request, selection, and initiate basis. A memory request or selection and a memory word 6-30

*Read only memory (ROM)*: Often, today, devices requiring 8 or more input variables are implemented using a ROM. A simple type of ROM can be constructed from a decoder, a MUX, and a number of wires. We will look at a small (16 bit) ROM constructed in this way. Normally, memory is arranged in a square array, as shown in Figure below. To use the ROM to implement a logic function, the address lines are used as the variable inputs, and the contents of the memory are the function values. Usually the memory has a word length of more than 1 bit; typically 4 or 8 bits, so several functions can be implemented simultaneously.

In Figure below, it is assumed that the decoder produces a logic 1 as output when the input code selects that output; otherwise it produces logic 0, and that a logic 1 output "outvotes" a logic 0 output, in the sense that if both are present on the same wire, the logic 1 will dominate. ("Real" circuits usually have the opposite behaviour). A bit is "programmed" when a link is present.

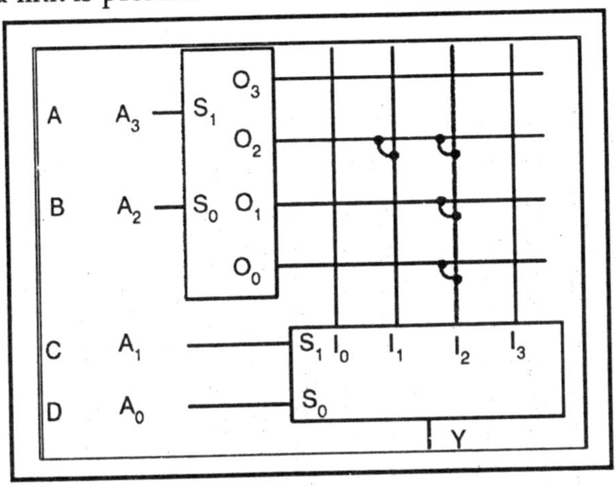


**Fig:** A read-only memory implementation

In the example of Figure above, the function is,

$$Y = \overline{A}\cdot\overline{B}\cdot C\cdot\overline{D} + \overline{A}\cdot B\cdot C\cdot\overline{D} + A\cdot\overline{B}\cdot\overline{C}\cdot D + A\cdot\overline{B}\cdot C\cdot\overline{D}$$

corresponding to memory locations 0010, 0110, 1001, and 1010.

This general organization is used for other types of memory, as well. One type of programmable read-only memory uses small "fuse links" to connect the horizontal and vertical wires at each intersection. This device is "programmed" by passing sufficient current through a link to "blow" the fuse. The link could also be a transistor which could be turned on or off, allowing a read-write type of memory to be implemented.

## DIGITAL STORAGE

### MEMORY

The basic goal of digital memory is to provide a means to store and access binary data: sequences of 1's and 0's. The digital storage of information holds advantages over analog techniques much the same as digital communication of information holds advantages over analog communication. This is not to say that digital data storage is unequivocally superior to analog, but it does address some of the more common problems associated with analog techniques and thus finds immense popularity in both consumer and industrial applications.

Digital data storage also complements digital computation technology well, and thus finds natural application in the world of computers. The most evident advantage of digital data storage is the resistance to corruption. Suppose that we were going to store a piece of data regarding the magnitude of a voltage signal by means of magnetizing a small chunk of magnetic material.

Since many magnetic materials retain their strength of magnetization very well over time, this would be a logical media candidate for long-term storage of this particular data (in fact, this is precisely how audio and video tape technology works: thin plastic tape is impregnated with particles of iron-oxide material, which can be magnetized or demagnetized via the application of a magnetic field from an electromagnet coil. The data is then retrieved from the tape by moving the magnetized tape past another coil of wire, the magnetized spots on the tape inducing voltage in that coil, reproducing the voltage waveform initially used to magnetize the tape).

If we represent an analog signal by the strength of magnetization on spots of the tape, the storage of data on the tape will be susceptible to the smallest degree of degradation of that magnetization. As the tape ages and the magnetization fades, the analog signal magnitude represented on the tape will appear to be less than what it was when we first recorded the data.

Also, if any spurious magnetic fields happen to alter the magnetization on the tape, even if it's only by a small amount, that altering of field strength will be interpreted upon re-play as an altering (or corruption) of the signal that was recorded. Since analog signals have infinite resolution, the smallest degree of change will have an impact on the integrity of the data storage.

If we were to use that same tape and store the data in binary digital form, however, the strength of magnetization on the tape would fall into two discrete levels: "high" and "low," with no valid in-between states. As the tape aged or was exposed to spurious magnetic fields, those same locations on the tape would experience slight alteration of magnetic field strength, but unless the alterations were extreme, no data corruption would occur upon re-play of the tape.

By reducing the resolution of the signal impressed upon the magnetic tape, we've gained significant immunity to the kind of degradation and "noise" typically plaguing stored analog data. On the other hand, our data resolution would be limited to the scanning rate and the number of bits output by the A/D converter which interpreted the original analog signal, so the reproduction wouldn't necessarily be "better" than with analog, merely more rugged. With the advanced technology of modern A/D's, though, the tradeoff is acceptable for most applications.

Also, by encoding different types of data into specific binary number schemes, digital storage allows us to archive a wide variety of information that is often difficult to encode in analog form. Text, for example, is represented quite easily with the binary ASCII code, seven bits for each character, including punctuation marks, spaces, and carriage returns. A wider range of text is encoded using the Unicode standard, in like manner.

Any kind of numerical data can be represented using binary notation on digital media, and any kind of information that can be encoded in numerical form (which almost any kind can!) is storable, too. Techniques such as parity and checksum error detection can be employed to further guard against data corruption, in ways that analog does not lend itself to.

## Concepts

When we store information in some kind of circuit or device, we not only need some way to store and retrieve it, but also to locate precisely where in the device that it is. Most, if not all, memory devices can be thought of as a series of mail boxes, folders in a file cabinet, or some other metaphor where information can be located in a variety of places. When we refer to the actual information being stored in the memory device, we usually refer to it as the data.

The location of this data within the storage device is typically called the address, in a manner reminiscent of the postal service.

With some types of memory devices, the address in which certain data is stored can be called up by means of parallel data lines in a digital circuit (we'll discuss this in more detail later in this lesson). With other types of devices, data is addressed in terms of an actual physical location on the surface of some type of media (the tracks and sectors of circular computer disks, for instance).

However, some memory devices such as magnetic tapes have a one-dimensional type of data addressing: if you want to play your favourite song in the middle of a cassette tape album, you have to fast-forward to that spot in the tape, arriving at the proper spot by means of trial-and-error, judging the approximate area by means of a counter that keeps track of tape position, and/or by the amount of time it takes to get there from the beginning of the tape.

The access of data from a storage device falls roughly into two categories: random access and sequential access. Random access means that you can quickly and precisely address a specific data location within the device, and non-random simply means that you cannot. A vinyl record platter is an example of a random-access device: to skip to any song, you just position the stylus arm at whatever location on the record that you want (compact audio disks so the same thing, only they do it automatically for you).

Cassette tape, on the other hand, is sequential. You have to wait to go past the other songs in sequence before you can access or address the song that you want to skip to. The process of storing a piece of data to a memory device is called writing, and the process of retrieving data is calledreading. Memory devices allowing both reading and writing are equipped with a way to distinguish between the two tasks, so that no mistake is made by the user (writing new information to a device when all you wanted to do is see what was stored there).

Some devices do not allow for the writing of new data, and are purchased "pre-written" from the manufacturer. Such is the case for vinyl records and compact audio disks, and this is typically referred to in the digital world as read-only memory, or ROM. Cassette audio and video tape, on the other hand, can be re-recorded (re-written) or purchased blank and recorded fresh by the user. This is often called read-write memory.

Another distinction to be made for any particular memory technology is its volatility, or data storage permanence without power. Many electronic memory devices store binary data by means of circuits that are either latched in a "high" or "low" state, and this latching effect holds only as long as electric power is maintained to those circuits. Such memory would be properly referred to as volatile. Storage media such as magnetized disk or tape is nonvolatile, because no source of power is needed to maintain data storage.

This is often confusing for new students of computer technology, because the volatile electronic memory typically used for the construction of computer devices is commonly and distinctly referred to as RAM(Random Access Memory). While RAM memory is typically randomly-accessed, so is virtually every other kind of memory device in the computer! What "RAM" really refers to is the volatility of the memory, and not its mode of access.

Nonvolatile memory integrated circuits in personal computers are commonly (and properly) referred to as ROM (Read-Only Memory), but their data contents are accessed randomly, just like the volatile memory circuits! Finally, there needs to be a way to denote how much data can be stored by any particular memory device. This, fortunately for us, is very simple and straightforward: just count up the number of bits (or bytes, 1 byte = 8 bits) of total data storage space.

Due to the high capacity of modern data storage devices, metric prefixes are generally affixed to the unit of bytes in order to represent storage space: 1.6 Gigabytes is equal to 1.6 billion bytes, or 12.8 billion bits, of data storage capacity. The only caveat here is to be aware of rounded numbers. Because the storage mechanisms of many random-access memory devices are typically arranged so that the number of "cells" in which bits of data can be stored appears in binary progression (powers of 2), a "one kilobyte" memory device most likely contains 1024 (2 to the power of 10) locations for data bytes rather than exactly 1000.

A "64 kbyte" memory device actually holds 65,536 bytes of data (2 to the 16th power), and should probably be called a "66 Kbyte" device to be more precise. When we round numbers in our base-10 system, we fall out of step with the round equivalents in the base-2 system.

## MODERN NONMECHANICAL MEMORY

Now we can proceed to studying specific types of digital storage devices. To start, I want to explore some of the technologies which do not require any moving parts. These are not necessarily the newest technologies, as one might suspect, although they will most likely replace moving-part technologies in the future.

A very simple type of electronic memory is the bistable multivibrator. Capable of storing a single bit of data, it is volatile (requiring power to maintain its memory) and very fast. The D-latch is probably the simplest implementation of a bistable multivibrator for memory usage, the D input serving as the data "write" input, the Q output serving as the "read" output, and the enable input serving as the read/write control line:

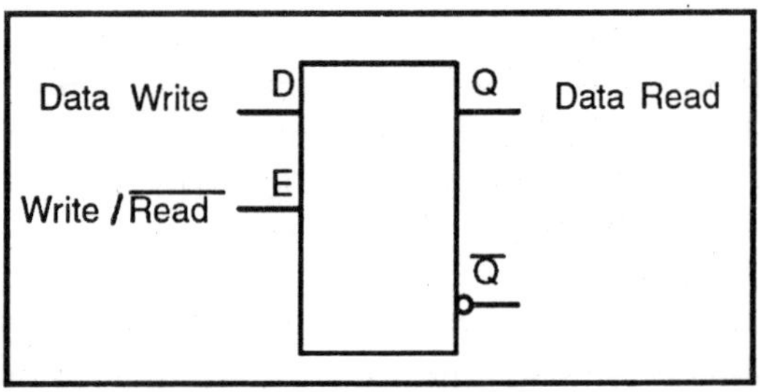

If we desire more than one bit's worth of storage (and we probably do), we'll have to have many latches arranged in some kind of an array where we can selectively address which one (or which set) we're reading from or writing to. Using a pair of tristate buffers, we can connect both the data write input and the data read output to a common data bus line, and enable those buffers to either connect the Q output to the data line (READ), connect the D input to the data line (WRITE), or keep both buffers in the High-Z state to disconnect D and Q from the data line (unaddressed mode).

This simple memory circuit is random-access and volatile. Technically, it is known as a static RAM. Its total memory capacity is 16 bits. Since it contains 16 addresses and has a data bus that is 1 bit wide, it would be designated as a 16 × 1 bit static RAM circuit.

This makes the static RAM a relatively low-density device, with less capacity than most other types of RAM technology per unit IC chip space. Because each cell circuit consumes a certain amount of power, the overall power consumption for a large array of cells can be quite high.

Early static RAM banks in personal computers consumed a fair amount of power and generated a lot of heat, too. CMOS IC technology has made it possible to lower the specific power consumption of static RAM circuits, but low storage density is still an issue. To address this, engineers turned to the capacitor instead of the bistable multivibrator as a means of storing binary data.

A tiny capacitor could serve as a memory cell, complete with a single MOSFET transistor for connecting it to the data bus for charging (writing a 1), discharging (writing a 0), or reading. Unfortunately, such tiny capacitors have very small capacitances, and their charge tends to "leak" away through any circuit impedances quite rapidly.

To combat this tendency, engineers designed circuits internal to the RAM memory chip which would periodically read all cells and recharge (or "refresh") the capacitors as needed. Although this added to the complexity of the circuit, it still required far less componentry than a RAM built of multivibrators. They called this type of memory circuit a dynamic RAM, because of its need of periodic refreshing.

Recent advances in IC chip manufacturing has led to the introduction of flash memory, which works on a capacitive storage principle like the dynamic RAM, but uses the insulated gate of a MOSFET as the capacitor itself.

Before the advent of transistors (especially the MOSFET), engineers had to implement digital circuitry with gates constructed from vacuum tubes. As you can imagine, the enormous comparative size and power consumption of a vacuum tube as compared to a transistor made memory circuits like static and dynamic RAM a practical impossibility. Other, rather ingenious, techniques to store digital data without the use of moving parts were developed.

## MEMORY TECHNOLOGIES

Perhaps the most ingenious technique was that of the delay line. A delay line is any kind of device which delays the propagation of a pulse or wave signal. If you've ever heard a sound echo back and forth through a canyon or cave, you've experienced an audio delay line: the noise wave travels at the speed of sound, bouncing off of walls and reversing direction of travel.

The delay line "stores" data on a very temporary basis if the signal is not strengthened periodically, but the very fact that it stores data at all is a phenomenon exploitable for memory technology. Early computer delay lines used long tubes filled with liquid mercury, which was used as the physical medium through which sound waves traveled along the length of the tube. An electrical/sound transducer was mounted at each end, one to create sound waves from electrical impulses, and the other to generate electrical impulses from sound waves.

The delay line concept suffered numerous limitations from the materials and technology that were then available. The EDVAC computer of the early 1950's used 128 mercury-filled tubes, each one about 5 feet long and storing a maximum of 384 bits. Temperature changes would affect the speed of sound in the mercury, thus skewing the time delay in each tube and causing timing problems.

Later designs replaced the liquid mercury medium with solid rods of glass, quartz, or special metal that delayed torsional (twisting) waves rather than longitudinal (lengthwise) waves, and operated at much higher frequencies. One such delay line used a special nickel-iron-titanium wire (chosen for its good temperature stability) about 95 feet in length, coiled to reduce the overall package size.

The total delay time from one end of the wire to the other was about 9.8 milliseconds, and the highest practical clock frequency was 1 MHz. This meant that approximately 9800 bits of data could be stored in the delay line wire at any given time. Given different means of delaying signals which wouldn't be so susceptible to environmental variables (such as serial pulses of light within a long optical fibre), this approach might someday find re-application.

Another approach experimented with by early computer engineers was the use of a cathode ray tube (CRT), the type commonly used for oscilloscope, radar, and television viewscreens, to store binary data. Normally, the focused and directed electron beam in a CRT would be used to make bits of phosphor chemical on the inside of the tube glow, thus producing a viewable image on the screen. In this application, however, the desired result was the creation of an electric charge on the glass of the screen by the impact of the electron beam, which would then be detected by a metal grid placed directly in front of the CRT.

Like the delay line, the so-called Williams Tube memory needed to be periodically refreshed with external circuitry to retain its data. Unlike the delay line mechanisms, it was virtually immune to the environmental factors of temperature and vibration. The IBM model 701 computer sported a Williams Tube memory with 4 Kilobyte capacity and a bad habit of "overcharging" bits on the tube screen with successive re-writes so that false "1" states might overflow to adjacent spots on the screen.

In other words, each core was addressed by the

intersection of row and column. The distinction between "set" and "reset" was the direction of thee core's magnetic polarity, and that bit value of data would be determined by the polarity of the voltages (with respect to ground) that the row and column wires would be energized.

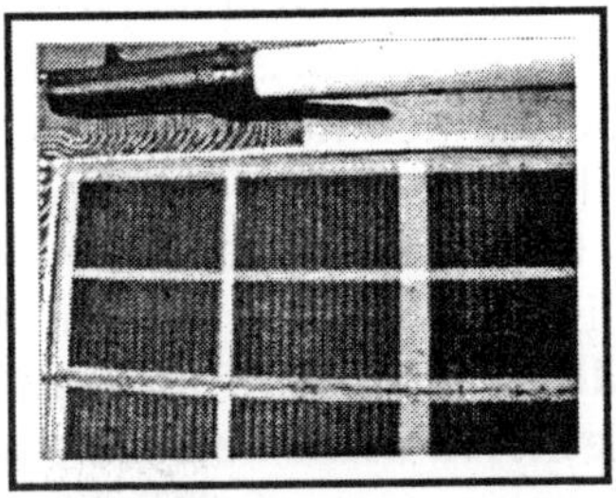

**Fig.** Another close-up of the cores:

Writing data to core memory was easy enough, but reading that data was a bit of a trick. To facilitate this essential function, a "read" wire was threaded through all the cores in a memory matrix, one end of it being grounded and the other end connected to an amplifier circuit. A pulse of voltage would be generated on this "read" wire if the addressed corechanged states (from 0 to 1, or 1 to 0).

In other words, to read a core's value, you had to write either a 1 or a 0 to that core and monitor the voltage induced on the read wire to see if the core changed. Obviously, if the core's state was changed, you would have to re-set it back to its original state, or else the data would have been lost.

This process is known as adestructive read, because data may be changed (destroyed) as it is read. Thus, refreshing is necessary with core memory, although not in every case (that is, in the case of the core's state not changing when either a 1 or a 0 was written to it).

One major advantage of core memory over delay lines and Williams Tubes was nonvolatility. The ferrite cores maintained their magnetization indefinitely, with no power or refreshing required. It was also relatively easy to build, denser, and physically more rugged than any of its predecessors. Core memory was used from the 1960's until the late 1970's in many computer systems, including the computers used for the gate-controlled MOSFET transistors acting as on/off switches to switch electrical signals, analog or digital. The "on" resistance of such a switch is in the range of several hundred ohms, the "off" resistance being in the range of several hundred mega-ohms.

Bilateral switches appear in schematics as SPST (Single-Pole, Single-Throw) switches inside of rectangular boxes, with a control terminal on one of the box's long sides:

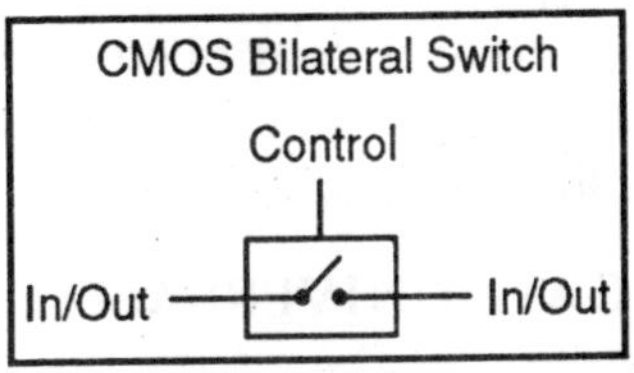


A bilateral switch might be best envisioned as a solid-state (semiconductor) version of an electromechanical relay: a signal-actuated switch contact that may be used to conduct virtually any type of electric signal. Of course, being solid-state, the bilateral switch has none of the undesirable characteristics of electromechanical relays, such as contact "bouncing," arcing, slow speed, or susceptibility to mechanical vibration. Conversely, though, they are rather limited in their current-carrying ability. Additionally, the signal conducted by the "contact" must not exceed the power supply "rail" voltages powering the bilateral switch circuit. Four bilateral switches are packaged inside the popular model "4066" integrated circuit:

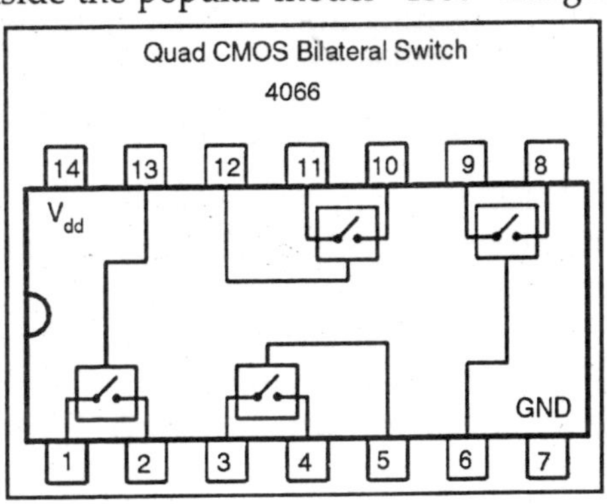


*Review*:

- Complementary gates provide both inverted and

noninverted output signals, in such a way that neither one is delayed with respect to the other.

- Tristate gates provide three different output states: high, low, and floating (High-Z). Such gates are commanded into their high-impedance output modes by a separate input terminal called the enable.
- Bilateral switches are MOSFET circuits providing on/off switching for a variety of electrical signal types (analog and digital), controlled by logic level voltage signals. In essence, they are solid-state relays with very low current-handling ability.

## GATE UNIVERSALITY

NAND and NOR gates possess a special property: they are universal. That is, given enough gates, either type of gate is able to mimic the operation of any other gate type. For example, it is possible to build a circuit exhibiting the OR function using three interconnected NAND gates.

The ability for a single gate type to be able to mimic any other gate type is one enjoyed only by the NAND and the NOR. In fact, digital control systems have been designed around nothing but either NAND or NOR gates, all the necessary logic functions being derived from collections of interconnected NANDs or NORs.

As proof of this property, this section will be divided into subsections showing how all the basic gate types may be formed using only NANDs or only NORs.

## CONSTRUCTING THE NOT FUNCTION

As you can see, there are two ways to use a NAND gate as an inverter, and two ways to use a NOR gate as an inverter. Either method works, although connecting TTL inputs together increases the amount of current loading to the driving gate. For CMOS gates, common input terminals decreases the switching speed of the gate due to increased input capacitance. Inverters are the fundamental tool for transforming one type of logic function into another, and so there will be many inverters shown in the illustrations to follow.

In those diagrams, I will only show one method of inversion, and that will be where the unused NAND gate input is connected to +V (either $V_{cc}$ or $V_{dd}$, depending on whether the circuit is TTL or CMOS) and where the unused input for the NOR gate is connected to ground.

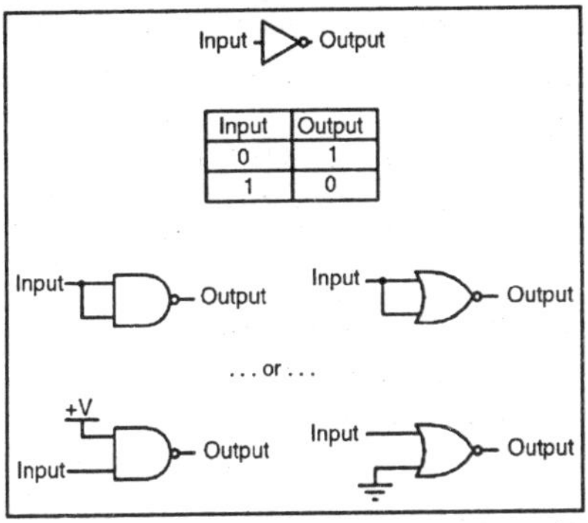


| Input | Output |
|---|---|
| 0 | 1 |
| 1 | 0 |

Bear in mind that the other inversion method (connecting both NAND or NOR inputs together) works just as well from a logical (1's and 0's) point of view, but is undesirable from the practical perspectives of increased current loading for TTL and increased input capacitance for CMOS.

## CONSTRUCTING THE "BUFFER" FUNCTION

Being that it is quite easy to employ NAND and NOR gates to perform the inverter (NOT) function, it stands to reason that two such stages of gates will result in a buffer function, where the output is the same logical state as the input.

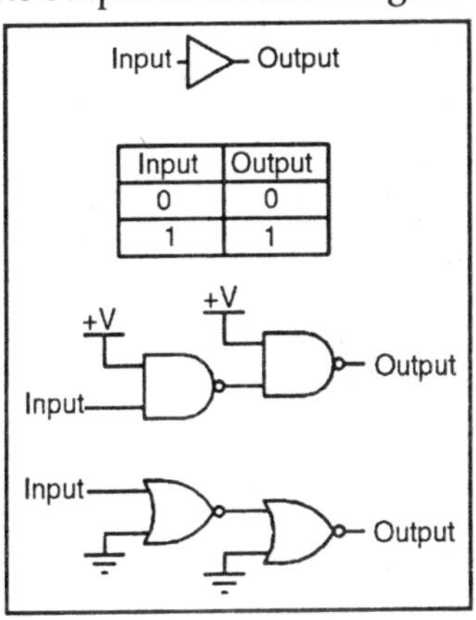


| Input | Output |
|---|---|
| 0 | 0 |
| 1 | 1 |

## CONSTRUCTING THE AND FUNCTION

To make the AND function from NAND gates, all that is needed is an inverter (NOT) stage on the output of a NAND gate. This extra inversion "cancels out" the first N in NAND, leaving the AND function. It takes a little more work to wrestle the same functionality out of NOR gates, but it can be done by inverting ("NOT") all of the inputs to a NOR gate.

2-Input AND Gate

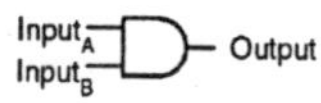


| A | B | Output |
|---|---|---|
| 0 | 0 | 0 |
| 0 | 1 | 0 |
| 1 | 0 | 0 |
| 1 | 1 | 1 |

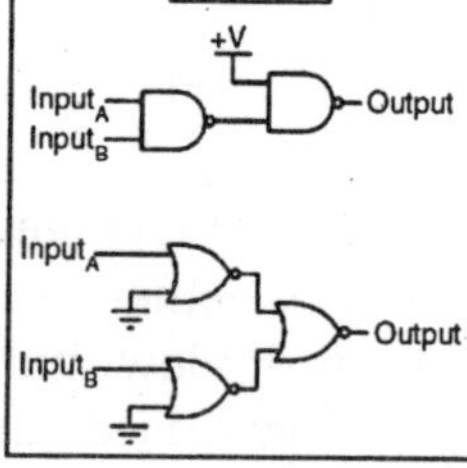


## CONSTRUCTING THE NAND FUNCTION

It would be pointless to show you how to "construct" the NAND function using a NAND gate, since there is nothing to do. To make a NOR gate perform the NAND function, we must invert all inputs to the NOR gate as well as the NOR gate's output. For a two-input gate, this requires three more NOR gates connected as inverters.

2-Input AND Gate

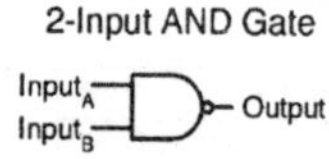


| A | B | Output |
|---|---|---|
| 0 | 0 | 1 |
| 0 | 1 | 1 |
| 1 | 0 | 1 |
| 1 | 1 | 0 |

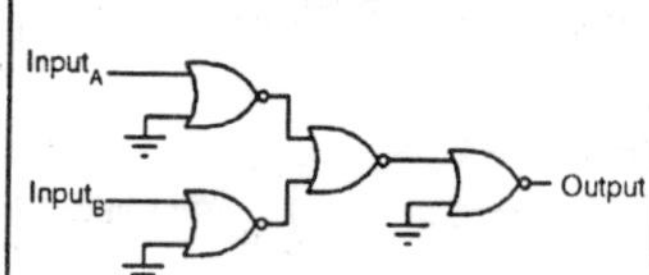

## CONSTRUCTING THE OR FUNCTION

Inverting the output of a NOR gate (with another NOR gate connected as an inverter) results in the OR function. The NAND gate, on the other hand, requires inversion of all inputs to mimic the OR function, just as we needed to invert all inputs of a NOR gate to obtain the AND function. Remember that inversion of all inputs to a gate results in changing that gate's essential function from AND to OR (or visa-versa), plus an inverted output.

Thus, with all inputs inverted, a NAND behaves as an OR, a NOR behaves as an AND, an AND behaves as a NOR, and an OR behaves as a NAND. In Boolean algebra, this transformation is referred to as DeMorgan's Theorem, covered in more detail in a later chapter of this book.

2-Input OR Gate

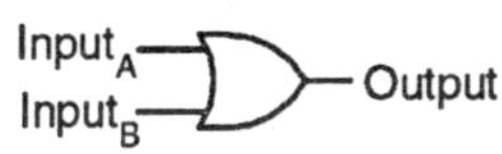


| A | B | Output |
|---|---|---|
| 0 | 0 | 0 |
| 0 | 1 | 1 |
| 1 | 0 | 1 |
| 1 | 1 | 1 |

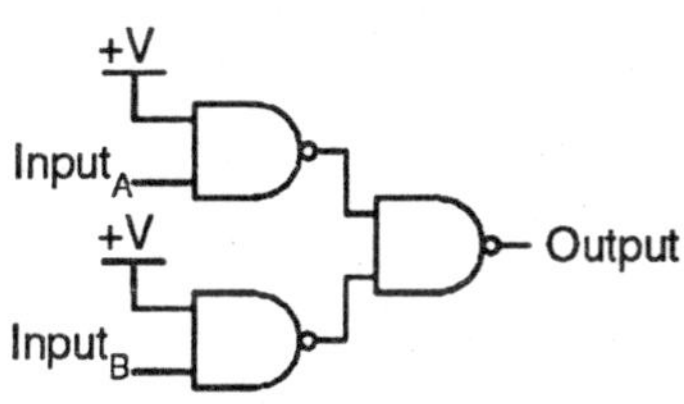


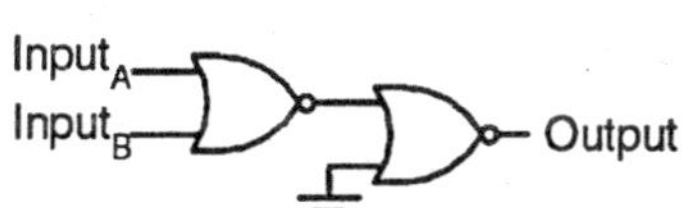


## CONSTRUCTING THE NOR FUNCTION

Much the same as the procedure for making a NOR gate

behave as a NAND, we must invert all inputs and the output to make a NAND gate function as a NOR.

2-Input NOR Gate

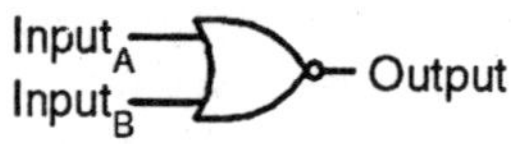


| A | B | Output |
|---|---|---|
| 0 | 0 | 1 |
| 0 | 1 | 0 |
| 1 | 0 | 0 |
| 1 | 1 | 0 |

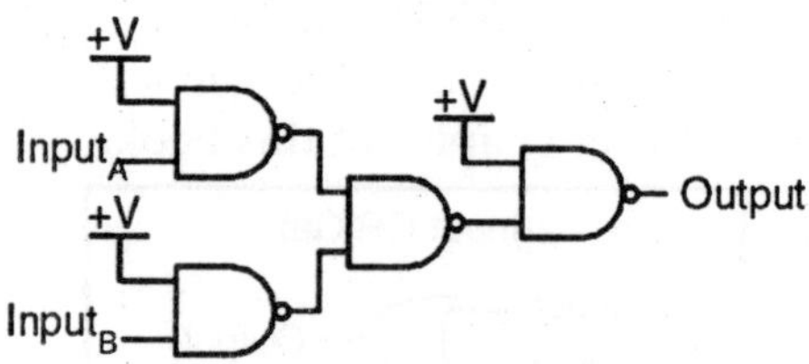


*Review*:

- NAND and NOR gates are universal: that is, they have the ability to mimic any type of gate, if interconnected in sufficient numbers.

## LOGIC SIGNAL VOLTAGE LEVELS

Logic gate circuits are designed to input and output only two types of signals: "high" (1) and "low" (0), as represented by a variable voltage: full power supply voltage for a "high" state and zero voltage for a "low" state.

In a perfect world, all logic circuit signals would exist at these extreme voltage limits, and never deviate from them (i.e., less than full voltage for a "high," or more than zero voltage for a "low").

However, in reality, logic signal voltage levels rarely attain these perfect limits due to stray voltage drops in the transistor circuitry, and so we must understand the signal level limitations of gate circuits as they try to interpret signal voltages lying somewhere between full supply voltage and zero.

TTL gates operate on a nominal power supply voltage of 5 volts, +/- 0.25 volts. Ideally, a TTL "high" signal would be 5.00 volts exactly, and a TTL "low" signal 0.00 volts exactly. However, real TTL gate circuits cannot output such perfect voltage levels, and are designed to accept "high" and "low" signals deviating substantially from these ideal values. "Acceptable" input signal voltages range from 0 volts to 0.8 volts for a "low" logic state, and 2 volts to 5 volts for a "high" logic state.

"Acceptable" output signal voltages (voltage levels guaranteed by the gate manufacturer over a specified range of load conditions) range from 0 volts to 0.5 volts for a "low" logic state, and 2.7 volts to 5 volts for a "high" logic state:

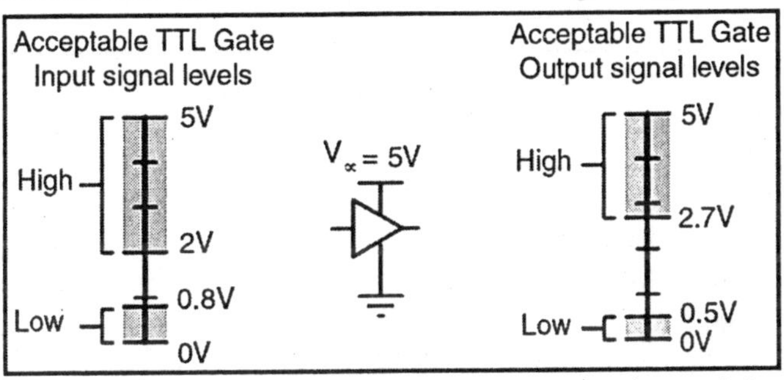


If a voltage signal ranging between 0.8 volts and 2 volts were to be sent into the input of a TTL gate, there would be no certain response from the gate.

Such a signal would be considered uncertain, and no logic gate manufacturer would guarantee how their gate circuit would interpret such a signal.

As you can see, the tolerable ranges for output signal levels are narrower than for input signal levels, to ensure that any TTL gate outputting a digital signal into the input of another TTL gate will transmit voltages acceptable to the receiving gate.

The difference between the tolerable output and input ranges is called the noise margin of the gate. For TTL gates, the low-level noise margin is the difference between 0.8 volts and 0.5 volts (0.3 volts), while the high-level noise margin is the difference between 2.7 volts and 2 volts (0.7 volts). Simply put, the noise margin is the peak amount of spurious or "noise" voltage that may be superimposed on a weak gate output voltage signal before the receiving gate might interpret it wrongly:

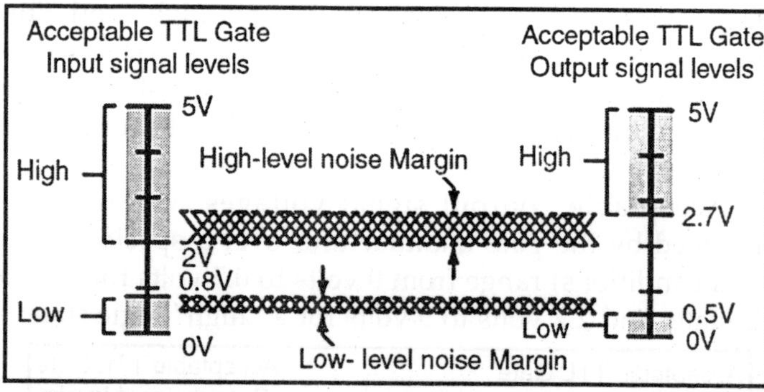


CMOS gate circuits have input and output signal specifications that are quite different from TTL. For a CMOS gate operating at a power supply voltage of 5 volts, the acceptable input signal voltages range from 0 volts to 1.5 volts for a "low" logic state, and 3.5 volts to 5 volts for a "high" logic state. "Acceptable" output signal voltages (voltage levels guaranteed by the gate manufacturer over a specified range of load conditions) range from 0 volts to 0.05 volts for a "low" logic state, and 4.95 volts to 5 volts for a "high" logic state:

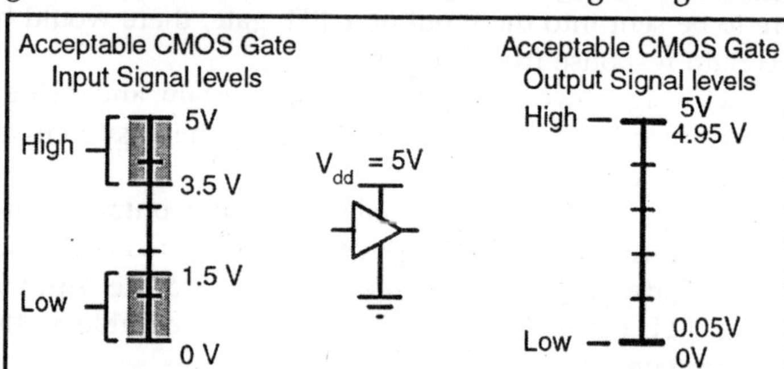


It should be obvious from these figures that CMOS gate circuits have far greater noise margins than TTL: 1.45 volts for CMOS low-level and high-level margins, versus a maximum of 0.7 volts for TTL. In other words, CMOS circuits can tolerate over twice the amount of superimposed "noise"

voltage on their input lines before signal interpretation errors will result.

CMOS noise margins widen even further with higher operating voltages. Unlike TTL, which is restricted to a power supply voltage of 5 volts, CMOS may be powered by voltages as high as 15 volts (some CMOS circuits as high as 18 volts). Shown here are the acceptable "high" and "low" states, for both input and output, of CMOS integrated circuits operating at 10 volts and 15 volts, respectively:

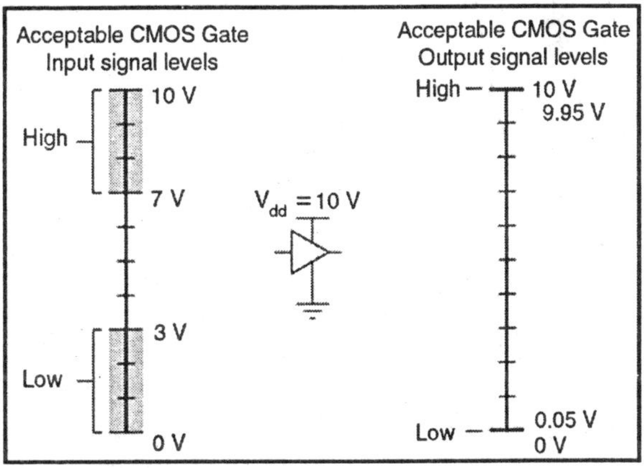


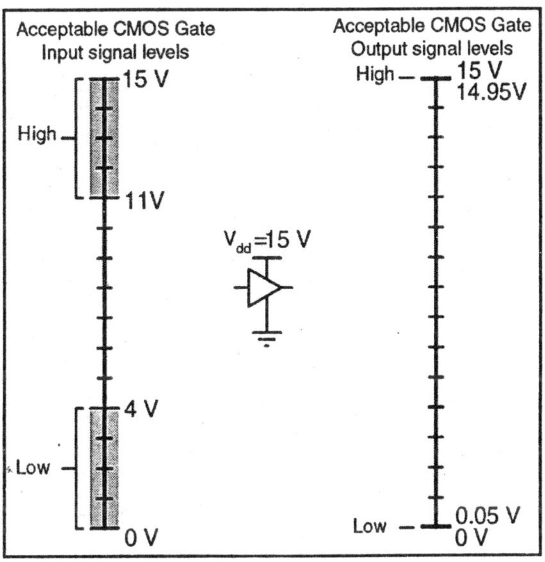

The margins for acceptable "high" and "low" signals may be greater than what is shown in the previous illustrations. What is shown represents "worst-case" input signal performance, based on manufacturer's specifications. In practice, it may be found that a gate circuit will tolerate "high" signals of considerably less voltage and "low" signals of considerably greater voltage than those specified here.

Conversely, the extremely small output margins shown — guaranteeing output states for "high" and "low" signals to within 0.05 volts of the power supply "rails" — are optimistic. Such "solid" output voltage levels will be true only for conditions of minimum loading. If the gate is sourcing or sinking substantial current to a load, the output voltage will not be able to maintain these optimum levels, due to internal channel resistance of the gate's final output MOSFETs.

Within the "uncertain" range for any gate input, there will be some point of demarcation dividing the gate's actual "low" input signal range from its actual "high" input signal range. That is, somewhere between the lowest "high" signal voltage level and the highest "low" signal voltage level guaranteed by the gate manufacturer, there is a threshold voltage at which the gate will actually switch its interpretation of a signal from "low" or "high" or visa-versa. For most gate circuits, this unspecified voltage is a single point:

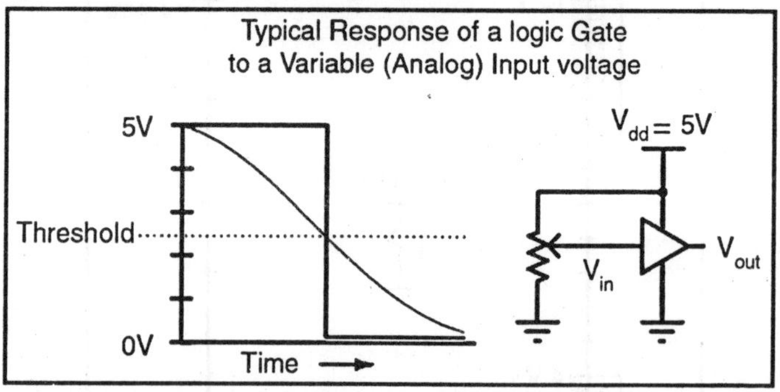


In the presence of AC "noise" voltage superimposed on the DC input signal, a single threshold point at which the gate alters its interpretation of logic level will result in an erratic output:

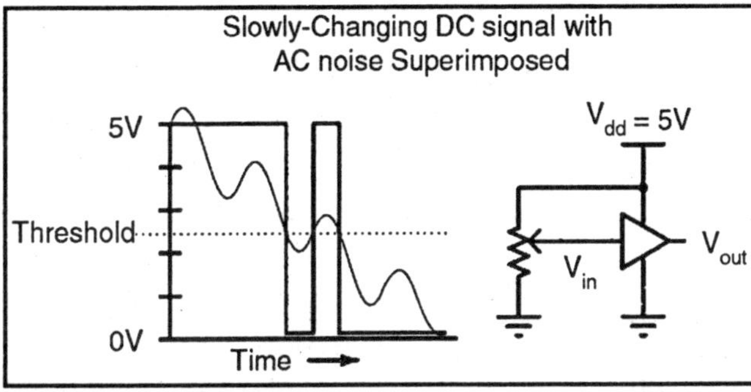


With a single threshold point at which an input causes the output to switch between "high" and "low" states, the presence of significant noise will cause erratic changes in the output:

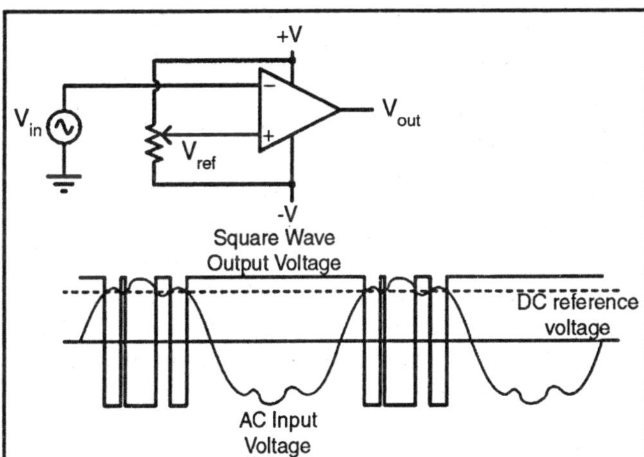


The solution to this problem is a bit of positive feedback introduced into the amplifier circuit. With an op-amp, this is done by connecting the output back around to the noninverting (+) input through a resistor.

In a gate circuit, this entails redesigning the internal gate circuitry, establishing the feedback inside the gate package rather than through external connections. A gate so designed is called a Schmitt trigger. Schmitt triggers interpret varying input voltages according to two threshold voltages: a positive-going threshold ($V_{T+}$), and a negative-going threshold ($V_{T-}$):

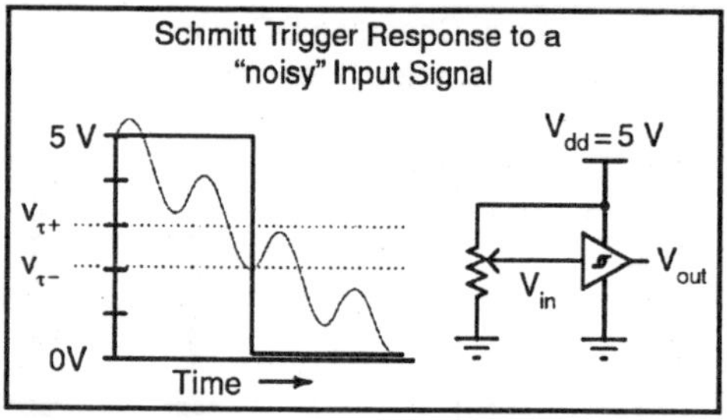


Schmitt trigger gates are distinguished in schematic diagrams by the small "hysteresis" symbol drawn within them, reminiscent of the B-H curve for a ferromagnetic material.

Schmitt trigger gates are frequently used in applications where noise is expected on the input signal line(s), and/or where an erratic output would be very detrimental to system performance. The differing voltage level requirements of TTL and CMOS technology present problems when the two types of gates are used in the same system.

Although operating CMOS gates on the same 5.00 volt power supply voltage required by the TTL gates is no problem, TTL output voltage levels will not be compatible with CMOS input voltage requirements. Take for instance a TTL NAND gate outputting a signal into the input of a CMOS inverter gate. Both gates are powered by the same 5.00 volt supply ($V_{cc}$). If the TTL gate outputs a "low" signal (guaranteed to be between 0 volts and 0.5 volts), it will be properly interpreted by the CMOS gate's input as a "low" (expecting a voltage between 0 volts and 1.5 volts):

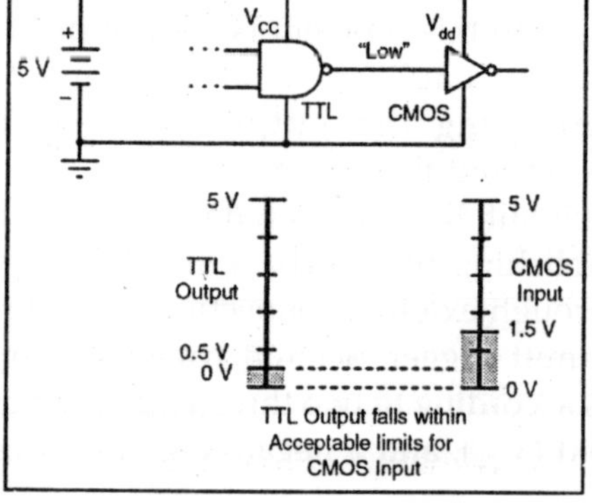

However, if the TTL gate outputs a "high" signal (guaranteed to be between 5 volts and 2.7 volts), it might not be properly interpreted by the CMOS gate's input as a "high" (expecting a voltage between 5 volts and 3.5 volts):

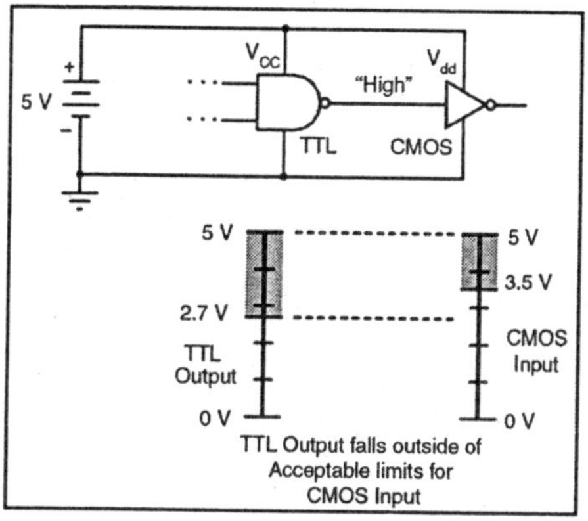


Given this mismatch, it is entirely possible for the TTL gate to output a valid "high" signal (valid, that is, according to the standards for TTL) that lies within the "uncertain" range for the CMOS input, and may be (falsely) interpreted as a "low" by the receiving gate. An easy "fix" for this problem is to augment the TTL gate's "high" signal voltage level by means of a pullup resistor:

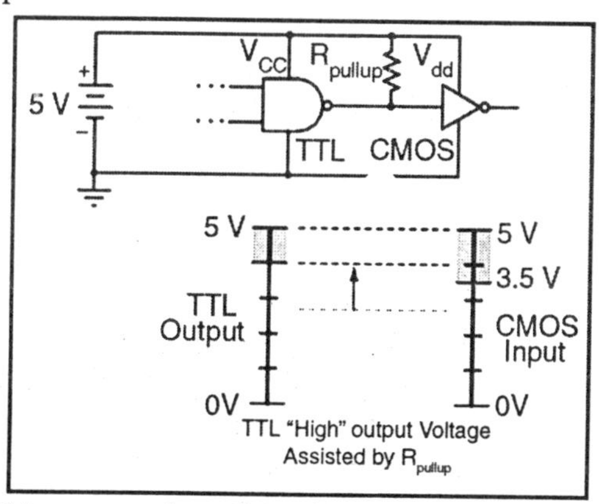


Something more than this, though, is required to interface a TTL output with a CMOS input, if the receiving CMOS gate is powered by a greater power supply voltage:

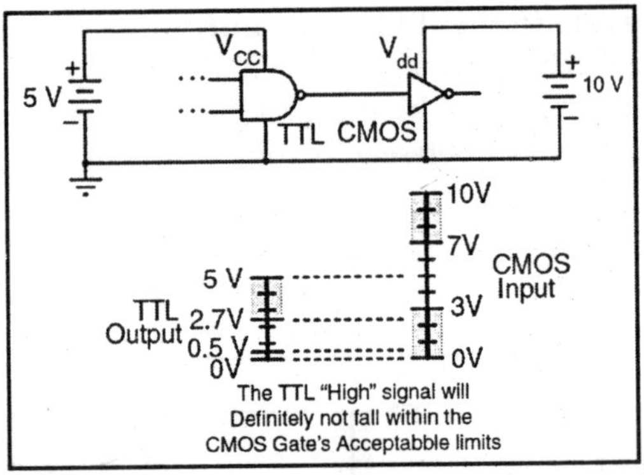


There will be no problem with the CMOS gate interpreting the TTL gate's "low" output, of course, but a "high" signal from the TTL gate is another matter entirely. The guaranteed output voltage range of 2.7 volts to 5 volts from the TTL gate output is nowhere near the CMOS gate's acceptable range of 7 volts to 10 volts for a "high" signal.

If we use an open-collector TTL gate instead of a totem-pole output gate, though, a pullup resistor to the 10 volt $V_{dd}$ supply rail will raise the TTL gate's "high" output voltage to the full power supply voltage supplying the CMOS gate. Since an open-collector gate can only sink current, not source current, the "high" state voltage level is entirely determined by the power supply to which the pullup resistor is attached, thus neatly solving the mismatch problem:

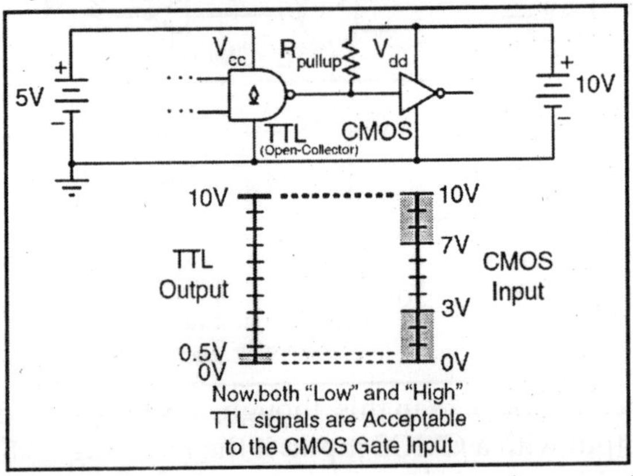


Due to the excellent output voltage characteristics of

CMOS gates, there is typically no problem connecting a CMOS output to a TTL input. The only significant issue is the current loading presented by the TTL inputs, since the CMOS output must sink current for each of the TTL inputs while in the "low" state.

When the CMOS gate in question is powered by a voltage source in excess of 5 volts ($V_{cc}$), though, a problem will result. The "high" output state of the CMOS gate, being greater than 5 volts, will exceed the TTL gate's acceptable input limits for a "high" signal. A solution to this problem is to create an "open-collector" inverter circuit using a discrete NPN transistor, and use it to interface the two gates together:

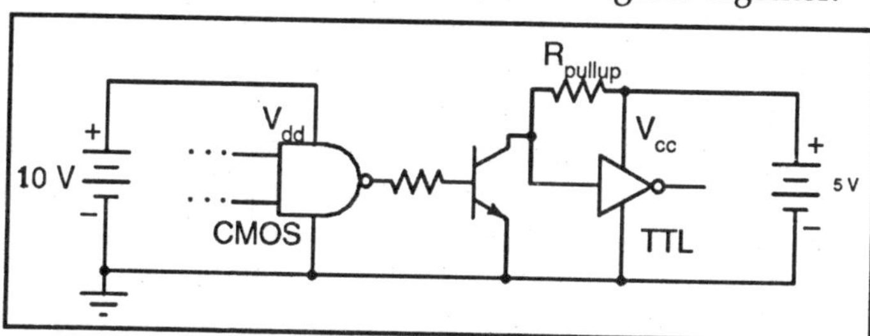


The "$R_{pullup}$" resistor is optional, since TTL inputs automatically assume a "high" state when left floating, which is what will happen when the CMOS gate output is "low" and the transistor cuts off. Of course, one very important consequence of implementing this solution is the logical inversion created by the transistor: when the CMOS gate outputs a "low" signal, the TTL gate sees a "high" input; and when the CMOS gate outputs a "high" signal, the transistor saturates and the TTL gate sees a "low" input. So long as this inversion is accounted for in the logical scheme of the system, all will be well.

## DIP GATE PACKAGING

Digital logic gate circuits are manufactured as integrated circuits: all the constituent transistors and resistors built on a single piece of semiconductor material. The engineer, technician, or hobbyist using small numbers of gates will likely find what he or she needs enclosed in a DIP (Dual Inline

Package) housing. DIP-enclosed integrated circuits are available with even numbers of pins, located at 0.100 inch intervals from each other for standard circuit board layout compatibility. Pin counts of 8, 14, 16, 18, and 24 are common for DIP "chips." Part numbers given to these DIP packages specify what type of gates are enclosed, and how many. These part numbers are industry standards, meaning that a "74LS02" manufactured by Motorola will be identical in function to a "74LS02" manufactured by Fairchild or by any other manufacturer. Letter codes prepended to the part number are unique to the manufacturer, and are not industry-standard codes. For instance, a SN74LS02 is a quad 2-input TTL NOR gate manufactured by Motorola, while a DM74LS02 is the exact same circuit manufactured by Fairchild.

Logic circuit part numbers beginning with "74" are commercial-grade TTL. If the part number begins with the number "54", the chip is a military-grade unit: having a greater operating temperature range, and typically more robust in regard to allowable power supply and signal voltage levels. The letters "LS" immediately following the 74/54 prefix indicate "Low-power Schottky" circuitry, using Schottky-barrier diodes and transistors throughout, to decrease power dissipation. Non-Schottky gate circuits consume more power, but are able to operate at higher frequencies due to their faster switching times. A few of the more common TTL "DIP" circuit packages are shown here for reference:

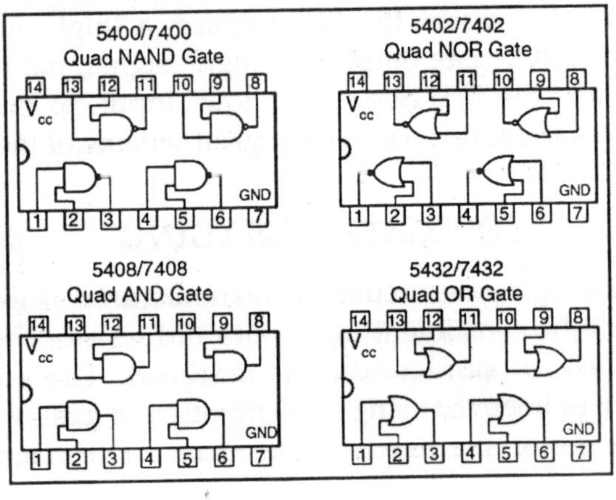

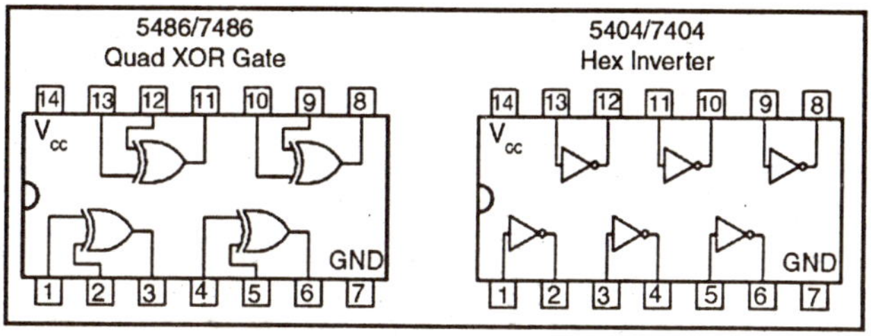
5486/7486
Quad XOR Gate
5404/7404
Hex Inverter
Vcc
GND
14
13
12
11
10
9
8
1
2
3
4
5
6
7

Chapter 5

# Combinational Logic Circuits

## BINARY ADDER

Suppose we wanted to build a device that could add two binary bits together. Such a device is known as a half-adder, and its gate circuit looks like this:

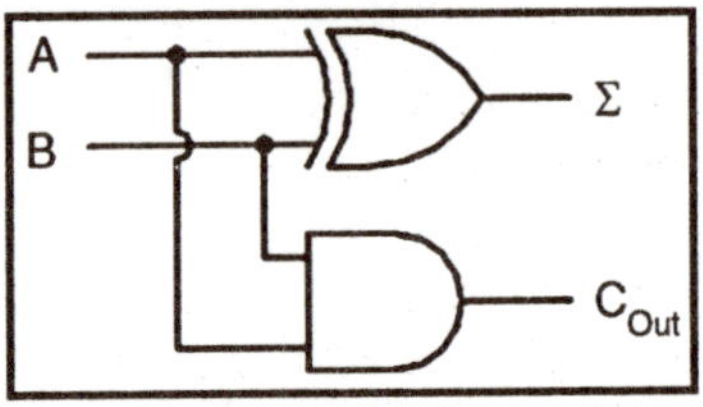


The Σ symbol represents the "sum" output of the half-adder, the sum's least significant bit (LSB). $C_{out}$ represents the "carry" output of the half-adder, the sum's most significant bit (MSB). If we were to implement this same function in ladder (relay) logic, it would look like this:

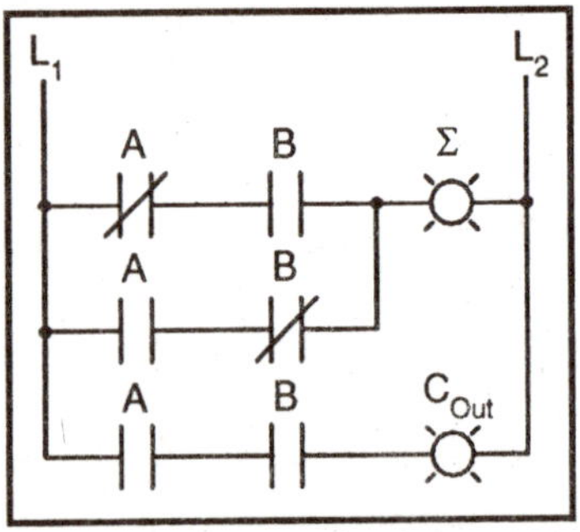


Either circuit is capable of adding two binary digits together. The mathematical "rules" of how to add bits together are intrinsic to the hard-wired logic of the circuits. If we wanted to perform a different arithmetic operation with binary bits, such as multiplication, we would have to construct another circuit.

The above circuit designs will only perform one function: add two binary bits together. To make them do something else would take re-wiring, and perhaps different compo entry.

In this sense, digital arithmetic circuits aren't much different from analog arithmetic (operational amplifier) circuits: they do exactly what they're wired to do, no more and no less.

We are not, however, restricted to designing digital computer circuits in this manner. It is possible to embed the mathematical "rules" for any arithmetic operation in the form of digital data rather than in hard-wired connections between gates.

The result is unparalleled flexibility in operation, giving rise to a whole new kind of digital device: the programmable computer.

## LOOK-UP TABLES

Having learned about digital memory devices in the last chapter, we know that it is possible to store binary data within solid-state devices.

Those storage "cells" within solid-state memory devices are easily addressed by driving the "address" lines of the device with the proper binary value(s).

Suppose we had a ROM memory circuit written, or programmed, with certain data, such that the address lines of the ROM served as inputs and the data lines of the ROM served as outputs, generating the characteristic response of a particular logic function.

Theoretically, we could programme this ROM chip to emulate whatever logic function we wanted without having to alter any wire connections or gates.

Consider the following example of a 4 × 2 bit ROM memory (a very small memory!) programmed with the functionality of a half adder:

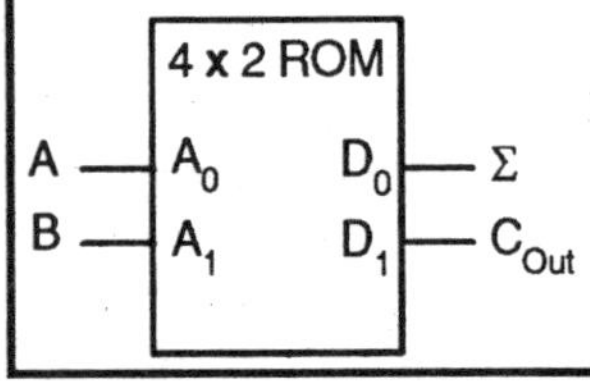


| Address | | Data | |
|---|---|---|---|
| A | B | $C_{Out}$ | Σ |
| 0 | 0 | 0 | 0 |
| 0 | 1 | 0 | 1 |
| 1 | 0 | 0 | 1 |
| 1 | 1 | 1 | 0 |

If this ROM has been written with the above data (representing a half-adder's truth table), driving the A and B address inputs will cause the respective memory cells in the ROM chip to be enabled, thus outputting the corresponding data as the Ó (Sum) and $C_{out}$ bits. Unlike the half-adder circuit built of gates or relays, this device can be set up to perform any logic function at all with two inputs and two outputs, not just the half-adder function. To change the logic function, all we would need to do is write a different table of data to another ROM chip. We could even use an EPROM chip which could be re-written at will, giving the ultimate flexibility in function.

It is vitally important to recognize the significance of this principle as applied to digital circuitry. Whereas the half-adder built from gates or relays processes the input bits to arrive at a specific output, the ROM simply remembers what the outputs should be for any given combination of inputs. This is not much different from the "times tables" memorized in grade school: rather than having to calculate the product of 5 times 6 (5 + 5 + 5 + 5 + 5 + 5 = 30), school-children are taught to remember that 5 × 6 = 30, and then expected to recall this product from memory as needed. Likewise, rather than the logic function depending on the functional arrangement of hard-wired gates or relays (hardware), it depends solely on the data written into the memory (software).

Such a simple application, with definite outputs for every input, is called a look-up table, because the memory device simply "looks up" what the output(s) should to be for any given combination of inputs states.

This application of a memory device to perform logical functions is significant for several reasons:

- Software is much easier to change than hardware.

- Software can be archived on various kinds of memory media (disk, tape), thus providing an easy way to document and manipulate the function in a "virtual" form; hardware can only be "archived" abstractly in the form of some kind of graphical drawing.
- Software can be copied from one memory device (such as the EPROM chip) to another, allowing the ability for one device to "learn" its function from another device.
- Software such as the logic function example can be designed to perform functions that would be extremely difficult to emulate with discrete logic gates (or relays!).

The usefulness of a look-up table becomes more and more evident with increasing complexity of function. Suppose we wanted to build a 4-bit adder circuit using a ROM. We'd require a ROM with 8 address lines (two 4-bit numbers to be added together), plus 4 data lines (for the signed output):

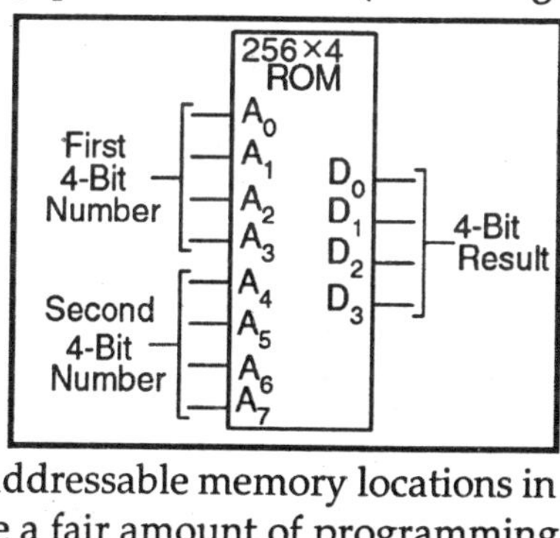


With 256 addressable memory locations in this ROM chip, we would have a fair amount of programming to do, telling it what binary output to generate for each and every combination of binary inputs.

We would also run the risk of making a mistake in our programming and have it output an incorrect sum, if we weren't careful. However, the flexibility of being able to configure this function (or any function) through software alone generally outweighs that costs.

Consider some of the advanced functions we could implement with the above "adder." We know that when we add two sets of numbers in 2's complement signed notation, we risk having the answer overflow. For instance, if we try to add 0111 (decimal 7) to 0110 (decimal 6) with only a 4-bit number field, the answer we'll get is 1001 (decimal -7) instead of the correct value, 13 (7 + 6), which cannot be expressed using 4 signed bits.

If we wanted to, we could avoid the strange answers given in overflow conditions by programming this look-up table circuit to output something else in conditions where we know overflow will occur (that is, in any case where the real sum would exceed +7 or -8).

One alternative might be to programme the ROM to output the quantity 0111 (the maximum positive value that can be represented with 4 signed bits), or any other value that we determined to be more appropriate for the application than the typical overflowed "error" value that a regular adder circuit would output.

It's all up to the programmer to decide what he or she wants this circuit to do, because we are no longer limited by the constraints of logic gate functions. The possibilities don't stop at customized logic functions, either. By adding more address lines to the 256 × 4 ROM chip, we can expand the look-up table to include multiple functions:

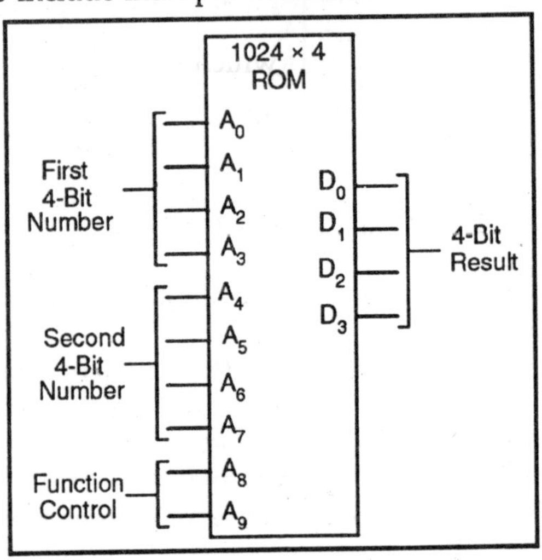

With two more address lines, the ROM chip will have 4 times as many addresses as before (1024 instead of 256). This ROM could be programmed so that when A8 and A9 were both low, the output data represented the sum of the two 4-bit binary numbers input on address lines A0 through A7, just as we had with the previous 256 × 4 ROM circuit.

For the addresses A8=1 and A9=0, it could be programmed to output the difference (subtraction) between the first 4-bit binary number (A0 through A3) and the second binary number (A4 through A7).

For the addresses A8=0 and A9=1, we could programme the ROM to output the difference (subtraction) of the two numbers in reverse order (second - first rather than first - second), and finally, for the addresses A8=1 and A9=1, the ROM could be programmed to compare the two inputs and output an indication of equality or inequality.

What we will have then is a device that can perform four different arithmetical operations on 4-bit binary numbers, all by "looking up" the answers programmed into it.

If we had used a ROM chip with more than two additional address lines, we could programme it with a wider variety of functions to perform on the two 4-bit inputs.

There are a number of operations peculiar to binary data (such as parity check or Exclusive-ORing of bits) that we might find useful to have programmed in such a look-up table.

Devices such as this, which can perform a variety of arithmetical tasks as dictated by a binary input code, are known as Arithmetic Logic Units (ALUs), and they comprise one of the essential components of computer technology.

Although modern ALUs are more often constructed from very complex combinational logic (gate) circuits for reasons of speed, it should be comforting to know that the exact same functionality may be duplicated with a "dumb" ROM chip programmed with the appropriate look-up table(s).

In fact, this exact approach was used by IBM engineers in 1959 with the development of the IBM 1401 and 1620 computers, which used look-up tables to perform addition, rather than binary adder circuitry.

The machine was fondly known as the "CADET," which stood for "Can't Add, Doesn't Even Try." A very common application for look-up table ROMs is in control systems where a custom mathematical function needs to be represented.

Such an application is found in computer-controlled fuel injection systems for automobile engines, where the proper air/fuel mixture ratio for efficient and clean operation changes with several environmental and operational variables. Tests performed on engines in research laboratories determine what these ideal ratios are for varying conditions of engine load, ambient air temperature, and barometric air pressure.

The variables are measured with sensor transducers, their analog outputs converted to digital signals with A/D circuitry, and those parallel digital signals used as address inputs to a high-capacity ROM chip programmed to output the optimum digital value for air/fuel ratio for any of these given conditions.

Sometimes, ROMs are used to provide one-dimensional look-up table functions, for "correcting" digitized signal values so that they more accurately represent their real-world significance. An example of such a device is a thermocouple transmitter, which measures the millivoltage signal generated by a junction of dissimilar metals and outputs a signal which is supposed to directly correspond to that junction temperature. Unfortunately, thermocouple junctions do not have perfectly linear temperature/voltage responses, and so the raw voltage signal is not perfectly proportional to temperature. By digitizing the voltage signal (A/D conversion) and sending that digital value to the address of a ROM programmed with the necessary correction values, the ROM's programming could eliminate some of the nonlinearity of the thermocouple's temperature-to-millivoltage relationship, so that the final output of the device would be more accurate. The popular instrumentation term for such a look-up table is a digital characterizer.

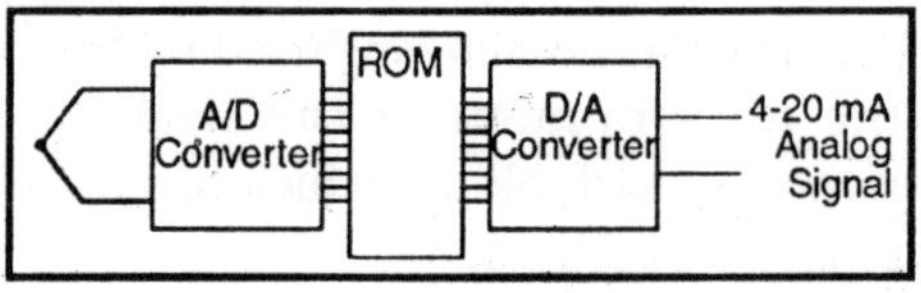

Another application for look-up tables is in special code translation. A 128 × 8 ROM, for instance, could be used to translate 7-bit ASCII code to 8-bit EBCDIC code:

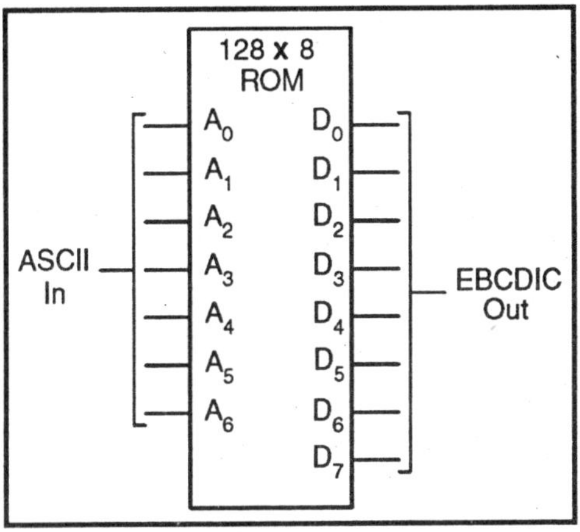


Again, all that is required is for the ROM chip to be properly programmed with the necessary data so that each valid ASCII input will produce a corresponding EBCDIC output code.

## FINITE-STATE MACHINES

Feedback is a fascinating engineering principle. It can turn a rather simple device or process into something substantially more complex. We've seen the effects of feedback intentionally integrated into circuit designs with some rather astounding effects:

- Comparator + negative feedback → controllable-gain amplifier
- Comparator + positive feedback → comparator with hysteresis
- Combinational logic + positive feedback → multivibrator

In the field of process instrumentation, feedback is used to transform a simple measurement system into something capable of control:

- Measurement system + negative feedback → closed-loop control system

Feedback, both positive and negative, has the tendency to add whole new dynamics to the operation of a device or system. Sometimes, these new dynamics find useful application, while other times they are merely interesting. With look-up tables programmed into memory devices, feedback from the data outputs back to the address inputs creates a whole new type of device: the Finite State Machine, or FSM:

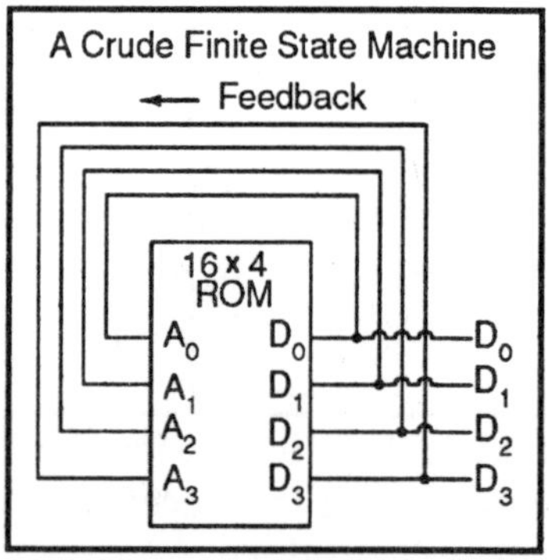


The above circuit illustrates the basic idea: the data stored at each address becomes the next storage location that the ROM gets addressed to. The result is a specific sequence of binary numbers (following the sequence programmed into the ROM) at the output, over time. To avoid signal timing problems, though, we need to connect the data outputs back to the address inputs through a 4-bit D-type flip-flop, so that the sequence takes place step by step to the beat of a controlled clock pulse:

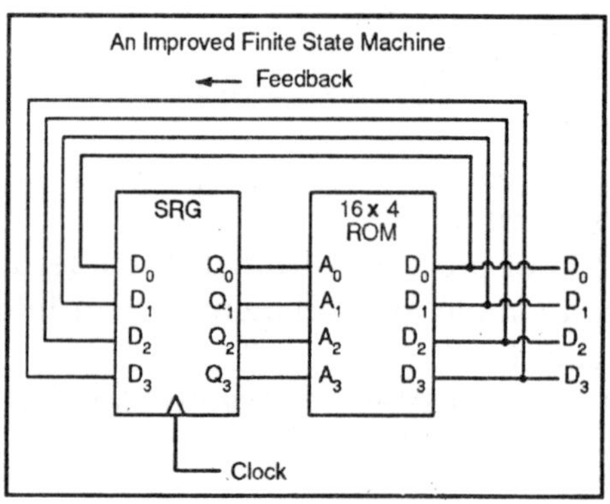

An analogy for the workings of such a device might be an array of post-office boxes, each one with an identifying number on the door (the address), and each one containing a piece of paper with the address of another P.O. box written on it (the data). A person, opening the first P.O. box, would find in it the address of the next P.O. box to open. By storing a particular pattern of addresses in the P.O. boxes, we can dictate the sequence in which each box gets opened, and therefore the sequence of which paper gets read.

Having 16 addressable memory locations in the ROM, this Finite State Machine would have 16 different stable "states" in which it could latch. In each of those states, the identity of the next state would be programmed in to the ROM, awaiting the signal of the next clock pulse to be fed back to the ROM as an address. One useful application of such an FSM would be to generate an arbitrary count sequence, such as Gray Code:

| Address | → | Data | Gray Code count sequence: | |
|---|---|---|---|---|
| 0000 | → | 0001 | 0 | 0000 |
| 0001 | → | 0011 | 1 | 0001 |
| 0010 | → | 0110 | 2 | 0011 |
| 0011 | → | 0010 | 3 | 0010 |
| 0100 | → | 1100 | 4 | 0110 |
| 0101 | → | 0100 | 5 | 0111 |
| 0110 | → | 0111 | 6 | 0101 |
| 0111 | → | 0101 | 7 | 0100 |
| 1000 | → | 0000 | 8 | 1100 |
| 1001 | → | 1000 | 9 | 1101 |
| 1010 | → | 1011 | 10 | 1111 |
| 1011 | → | 1001 | 11 | 1110 |
| 1100 | → | 1101 | 12 | 1010 |
| 1101 | → | 1111 | 13 | 1011 |
| 1110 | → | 1010 | 14 | 1001 |
| 1111 | → | 1110 | 15 | 1000 |

Try to follow the Gray Code count sequence as the FSM would do it: starting at 0000, follow the data stored at that address (0001) to the next address, and so on (0011), and so on (0010), and so on (0110), etc. The result, for the programme table shown, is that the sequence of addressing jumps around from address to address in what looks like a haphazard fashion, but when you check each address that is accessed, you will find that it follows the correct order for 4-bit Gray code. When the FSM arrives at its last programmed state (address 1000), the data stored there is 0000, which starts the whole sequence over again at address 0000 in step with the next clock pulse.

We could expand on the capabilities of the above circuit by using a ROM with more address lines, and adding more programming data:

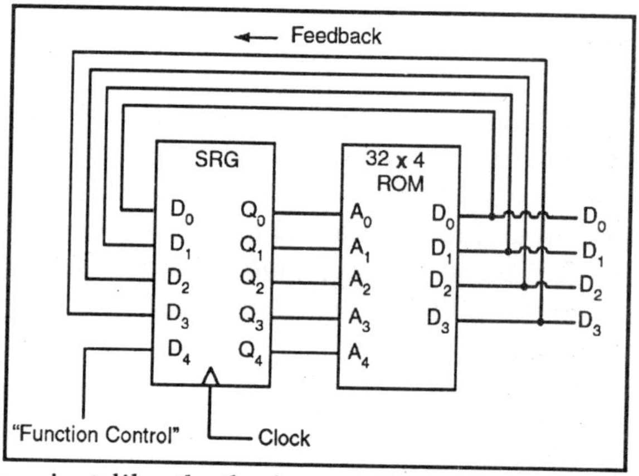


Now, just like the look-up table adder circuit that we turned into an Arithmetic Logic Unit (+, -, x, / functions) by utilizing more address lines as "function control" inputs, this FSM counter can be used to generate more than one count sequence, a different sequence programmed for the four feedback bits (A0 through A3) for each of the two function control line input combinations (A4 = 0 or 1).

| Address | → | Data | Address | → | Data |
|---|---|---|---|---|---|
| 00000 | → | 0001 | 10000 | → | 0001 |
| 00001 | → | 0010 | 10001 | → | 0011 |
| 00010 | → | 0011 | 10010 | → | 0110 |
| 00011 | → | 0100 | 10011 | → | 0010 |
| 00100 | → | 0101 | 10100 | → | 1100 |
| 00101 | → | 0110 | 10101 | → | 0100 |
| 00110 | → | 0111 | 10110 | → | 0111 |
| 00111 | → | 1000 | 10111 | → | 0101 |

```
01000  →  1001    11000  →  0000
01001  →  1010    11001  →  1000
01010  →  1011    11010  →  1011
01011  →  1100    11011  →  1001
01100  →  1101    11100  →  1101
01101  →  1110    11101  →  1111
01110  →  1111    11110  →  1010
01111  →  0000    11111  →  1110
```

If A4 is 0, the FSM counts in binary; if A4 is 1, the FSM counts in Gray Code. In either case, the counting sequence is arbitrary: determined by the whim of the programmer. For that matter, the counting sequence doesn't even have to have 16 steps, as the programmer may decide to have the sequence recycle to 0000 at any one of the steps at all. It is a completely flexible counting device, the behaviour strictly determined by the software (programming) in the ROM.

We can expand on the capabilities of the FSM even more by utilizing a ROM chip with additional address input and data output lines.

*Take the following circuit, for example:*

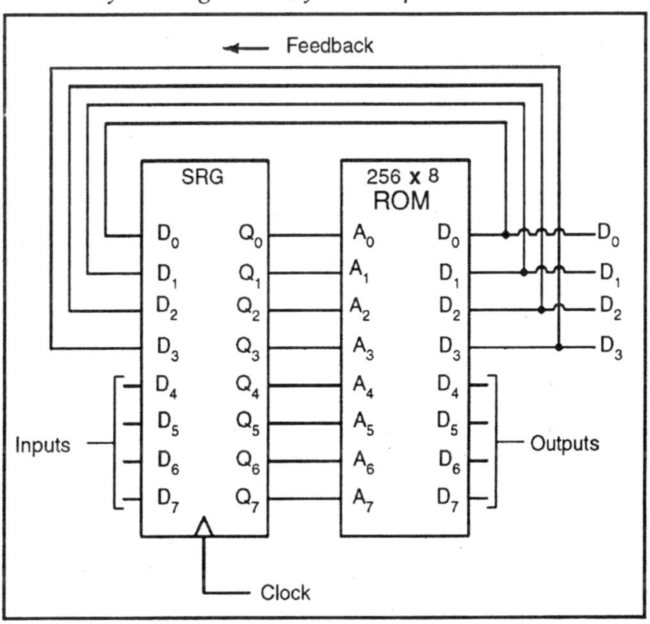

Here, the D0 through D3 data outputs are used exclusively for feedback to the A0 through A3 address lines. Date output lines D4 through D7 can be programmed to output something other than the FSM's "state" value. Being that four data output bits are being fed back to four address bits, this is still a 16-state device.

However, having the output data come from other data output lines gives the programmer more freedom to configure functions than before. In other words, this device can do far more than just count! The programmed output of this FSM is dependent not only upon the state of the feedback address lines (A0 through A3), but also the states of the input lines (A4 through A7).

The D-type flip/flop's clock signal input does not have to come from a pulse generator, either. To make things more interesting, the flip/flop could be wired up to clock on some external event, so that the FSM goes to the next state only when an input signal tells it to.

Now we have a device that better fulfills the meaning of the word "programmable." The data written to the ROM is a programme in the truest sense: the outputs follow a pre-established order based on the inputs to the device and which "step" the device is on in its sequence. This is very close to the operating design of the Turing Machine, a theoretical computing device invented by Alan Turing, mathematically proven to be able to solve any known arithmetic problem, given enough memory capacity.

## MICROPROCESSORS

Early computer science pioneers such as Alan Turing and John Von Neumann postulated that for a computing device to be really useful, it not only had to be able to generate specific outputs as dictated by programmed instructions, but it also had to be able to write data to memory, and be able to act on that data later. Both the programme steps and the processed data were to reside in a common memory "pool," thus giving way to the label of the stored-programme computer.

Turing's theoretical machine utilized a sequential-access

tape, which would store data for a control circuit to read, the control circuit re-writing data to the tape and/or moving the tape to a new position to read more data. Modern computers use random-access memory devices instead of sequential-access tapes to accomplish essentially the same thing, except with greater capability.

A helpful illustration is that of early automatic machine tool control technology. Called open-loop, or sometimes just NC (numerical control), these control systems would direct the motion of a machine tool such as a lathe or a mill by following instructions programmed as holes in paper tape. The tape would be run one direction through a "read" mechanism, and the machine would blindly follow the instructions on the tape without regard to any other conditions. While these devices eliminated the burden of having to have a human machinist direct every motion of the machine tool, it was limited in usefulness.

Because the machine was blind to the real world, only following the instructions written on the tape, it could not compensate for changing conditions such as expansion of the metal or wear of the mechanisms. Also, the tape programmer had to be acutely aware of the sequence of previous instructions in the machine's programme to avoid troublesome circumstances (such as telling the machine tool to move the drill bit laterally while it is still inserted into a hole in the work), since the device had no memory other than the tape itself, which was read-only.

Upgrading from a simple tape reader to a Finite State control design gave the device a sort of memory that could be used to keep track of what it had already done (through feedback of some of the data bits to the address bits), so at least the programmer could decide to have the circuit remember "states" that the machine tool could be in (such as "coolant on," or tool position). However, there was still room for improvement.

The ultimate approach is to have the programme give instructions which would include the writing of new data to a read/write (RAM) memory, which the programme could easily recall and process. This way, the control system could record what it had done, and any sensor-detectable process changes, much in the same way that a human machinist might jot down notes or measurements on a scratch-pad for future reference in his or her work. This is what is referred to as CNC, or Closed-loop Numerical Control.

Engineers and computer scientists looked forward to the possibility of building digital devices that could modify their own programming, much the same as the human brain adapts the strength of inter-neural connections depending on environmental experiences (that is why memory retention improves with repeated study, and behaviour is modified through consequential feedback). Only if the computer's programme were stored in the same writable memory "pool" as the data would this be practical. It is interesting to note that the notion of a self-modifying programme is still considered to be on the cutting edge of computer science. Most computer programming relies on rather fixed sequences of instructions, with a separate field of data being the only information that gets altered.

To facilitate the stored-programme approach, we require a device that is much more complex than the simple FSM, although many of the same principles apply. First, we need read/write memory that can be easily accessed: this is easy enough to do. Static or dynamic RAM chips do the job well, and are inexpensive.

Secondly, we need some form of logic to process the data stored in memory. Because standard and Boolean arithmetic functions are so useful, we can use an Arithmetic Logic Unit (ALU) such as the look-up table ROM example explored earlier. Finally, we need a device that controls how and where data flows between the memory, the ALU, and the outside world.

This so-called Control Unit is the most mysterious piece of the puzzle yet, being comprised of tri-state buffers (to direct data to and from buses) and decoding logic which interprets certain binary codes as instructions to carry out. Sample instructions might be something like: "add the number stored at memory address 0010 with the number stored at memory address 1101," or, "determine the parity of the data in memory address 0111."

The choice of which binary codes represent which instructions for the Control Unit to decode is largely arbitrary, just as the choice of which binary codes to use in representing the letters of the alphabet in the ASCII standard was largely arbitrary. ASCII, however, is now an internationally recognized standard, whereas control unit instruction codes are almost always manufacturer-specific.

Putting these components together (read/write memory, ALU, and control unit) results in a digital device that is typically called a processor. If minimal memory is used, and all the necessary components are contained on a single integrated circuit, it is called a microprocessor. When combined with the necessary bus-control support circuitry, it is known as a Central Processing Unit, or CPU.

CPU operation is summed up in the so-called fetch/execute cycle. Fetch means to read an instruction from memory for the Control Unit to decode. A small binary counter in the CPU (known as the programme counter or instruction pointer) holds the address value where the next instruction is stored in main memory.

The Control Unit sends this binary address value to the main memory's address lines, and the memory's data output is read by the Control Unit to send to another holding register. If the fetched instruction requires reading more data from memory (for example, in adding two numbers together, we have to read both the numbers that are to be added from main memory or from some other source), the Control Unit appropriately addresses the location of the requested data and directs the data output to ALU registers.

## MICROPROCESSOR PROGRAMMING

The "vocabulary" of instructions which any particular microprocessor chip possesses is specific to that model of chip. An Intel 80386, for example, uses a completely different set of binary codes than a Motorola 68020, for designating equivalent functions. Unfortunately, there are no standards in place for microprocessor instructions. This makes programming at the very lowest level very confusing and specialized.

When a human programmer develops a set of instructions to directly tell a microprocessor how to do something (like automatically control the fuel injection rate to an engine), they're programming in the CPU's own "language." This language, which consists of the very same binary codes which the Control Unit inside the CPU chip decodes to perform tasks, is often referred to as machine language. While machine language software can be "worded" in binary notation, it is often written in hexadecimal form, because it is easier for human beings to work with.

There are even programs written for computers called assemblers which understand these mnemonics, translating them to the appropriate binary codes for a specified target microprocessor, so that the programmer can write a programme in the computer's native language without ever having to deal with strange hex or tedious binary code notation.

Once a programme is developed by a person, it must be written into memory before a microprocessor can execute it. If the programme is to be stored in ROM (which some are), this can be done with a special machine called a ROM programmer, or (if you're masochistic), by plugging the ROM chip into a breadboard, powering it up with the appropriate voltages, and writing data by making the right wire connections to the address and data lines, one at a time, for each instruction.

If the programme is to be stored in volatile memory, such as the operating computer's RAM memory, there may be a way to type it in by hand through that computer's keyboard (some computers have a mini-programme stored in ROM which tells the microprocessor how to accept keystrokes from a keyboard and store them as commands in RAM), even if it is too dumb to do anything else.

Many "hobby" computer kits work like this. If the computer to be programmed is a fully-functional personal computer with an operating system, disk drives, and the whole works, you can simply command the assembler to store your finished programme onto a disk for later retrieval. To "run" your programme, you would simply type your program's filename at the prompt, press the Enter key, and the microprocessor's Programme Counter register would be set to point to the location ("address") on the disk where the first instruction is stored, and your programme would run from there.

On the other hand, if the entire programme is translated to machine language in one fell swoop, like a translator recording a monologue on paper and then translating all the words at one sitting into a written document in the other language, the process is called "compilation."

Interpretation is simple, but makes for a slow-running programme because the microprocessor has to continually translate the programme between steps, and that takes time. Compilation takes time initially to translate the whole programme into machine code, but the resulting machine code needs no translation after that and runs faster as a consequence.

Programming languages such as BASIC and FORTH are interpreted. Languages such as C, C++, FORTRAN, and PASCAL are compiled. Compiled languages are generally considered to be the languages of choice for professional programmers, because of the efficiency of the final product.

Naturally, because machine language vocabularies vary widely from microprocessor to microprocessor, and since high-level languages are designed to be as universal as possible, the interpreting and compiling programs necessary for language translation must be microprocessor-specific. Development of these interpreters and compilers is a most impressive feat: the people who make these programs most definitely earn their keep, especially when you consider the work they must do to keep their software product current with the rapidly-changing microprocessor models appearing on the market!

To mitigate this difficulty, the trend-setting manufacturers of microprocessor chips (most notably, Intel and Motorola) try to design their new products to be backwardly compatible with their older products. For example, the entire instruction set for the Intel 80386 chip is contained within the latest Pentium IV chips, although the Pentium chips have additional instructions that the 80386 chips lack.

What this means is that machine-language programs (compilers, too) written for 80386 computers will run on the latest and greatest Intel Pentium IV CPU, but machine-language programs written specifically to take advantage of the Pentium's larger instruction set will not run on an 80386, because the older CPU simply doesn't have some of those instructions in its vocabulary: the Control Unit inside the 80386 cannot decode them.

## Chapter 6

# Logic Families

### DIODE

A two-terminal semiconductor (rectifying) device that exhibits a nonlinear current-voltage characteristic. The function of a diode is to allow current in one direction and to block current in the opposite direction. The terminals of a diode are called the anode and cathode. There are two kinds of semiconductor diodes: a P-N junction diode, which forms an electrical barrier at the interface between N- and P-type semiconductor layers, and a Schottky diode, whose barrier is formed between metal and semiconductor regions.

Semiconductors are crystals that, in their pure state, are resistive (that is, their electrical properties lie between those of conductors and insulators) – but when the proper impurities are added (this process is called doping) in trace amounts (often measured in parts per billion), display interesting and useful properties.

### HISTORY

The oldest ancestor of semiconductor devices was the crystal detector, used in early wireless radios. This device (patented by a German scientist, Ferdinand Braun, in 1899) was made of a single metal wire (fondly called a "cat's whisker") touching against a semiconductor crystal. The result was a "rectifying diode" (so called because it has two terminals), which lets current through easily one way, but hinders flow the other way. By 1930, though, vacuum-tube diodes had all but replaced the smaller but much quirkier crystal detector. The crystal and "cat's whisker" were left to languish as a kids' toy in the form of "crystal radios."

The development of radar during World War II did much to revive the fortunes of crystal detectors (and, as a result, that of semiconductors) — although temperamental, crystals were better than vacuum-tube diodes at rectifying the high frequencies used by radar. So, during the war, much effort was put into improving the semiconductors, mostly silicon and germanium, used in crystal detectors.

At about the same time, Russell Ohl at Bell Laboratories discovered that these materials could be "doped" with small amounts of foreign "impurity" atoms to create interesting new properties. Depending on the selection of impurities (often called dopants) added, semiconductor material of two electricallly-different types can be created — one that is electron-rich (called N-type, where N stands for Negative), or one that is electron-poor (called P-type, where P stands for Positive).

Most of the "magic" of semiconductor devices occurs at the boundary between P-type and N-type semiconductor material — such a boundary is called a P-N junction. Ohl and his colleagues found that such a P-N junction made an effective diode. Like many components, diodes have a positive side or leg (a.k.a, their anode), and a negative side (cathode). When the voltage on the anode is higher than on the cathode then current flows through the diode (the resistance is very low).

When the voltage is lower on the anode than on the cathode then the current does not flow (the resistance is very high). An easy way to remember this is to look at the symbol for a diode — the "arrow" in the diode symbol points the direction in which it allows current (hole flow) to flow. The cathode of a diode is generally marked with a line next to it (on the diode body). You can see a similar line in the schematic symbols, above.

Diodes are also some times marked with an identifying colour code (similar, but not identical, to that used for resistors); a good explanation is given here. Note that when current is flowing through a diode, the voltage on the positive leg is higher than on the negative leg (this is often referred to as the diode's "forward voltage drop"). The magnitude of the voltage drop is a function of (among other things) the semiconductor material that the diode is made from.

Silicon diodes are the most common and cheapest, and have a forward voltage drop of about 0.65 volts. Germanium diodes have a forward voltage drop of about 0.1 volt. Germanium diodes, though, are typically much more expensive than silicon diodes; luckily, they're salvageable from lots of circuit boards.

## TYPES OF DIODES

### Rectifying Diodes

The principal early application of diodes was in rectifying 50 / 60 Hz AC mains to raw DC which was later smoothed by choke transformers and / or capacitors. This procedure is still carried out today and a number of rectifying schemes for diodes have evolved, half wave, full wave and bridge rectifiers.

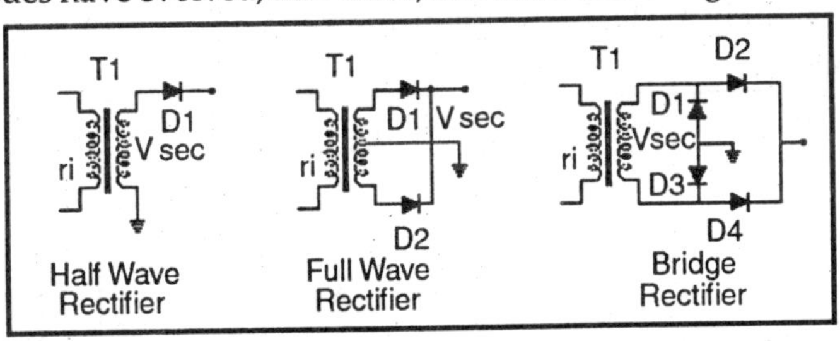


**Fig.** Rectifying Diodes.

As examples in these applications the half wave rectifier passes only the positive half of successive cycles to the output filter through D1. During the negative part of the cycle D1 does not conduct and no current flows to the load. In the full wave application it essentially is two half wave rectifiers combined and because the transformer secondary is centre tapped, D1 conducts on the positive half of the cycle while D2 conducts on the negative part of the cycle. Both add together. This is more efficient. The full wave bridge rectifier operates essentially the same as the full wave rectifier but does not require a cetre tapped transformer. Further discussion may be seen on the topic power supplies.

A further application of rectifying diodes is in the conversion or detection of rf modulated signals to audio frequencies. Typical examples are am modulated signals being detected and early detection schemes for fm also used diodes for detecting modulation.

## Voltage Regulating Diodes

For relatively light current loads zener diodes are a cheap solution to voltage regulation. Zener diodes work on the principle of essentially a constant voltage drop at a predetermined voltage (determined during manufacture). An example is a Philips BZX79C12 type with a regulation range between 11.4V and 12.7V but typically 12V and a total power dissipation of 500 mW in a DO-35 package. The dissipation can be extended by using a series pass transistor, see power supplies. Notice in figure below there is a resistor to miminmise current drawn but mainly as an aid to dropping the supply voltage and reducing the burden on the zener diodes.

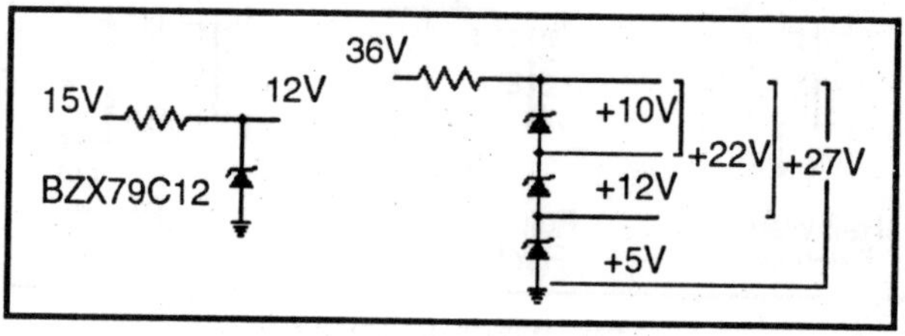


Figure 3 - zener voltage regulation diodes

In the second schematic of figure above we have three zener diodes in series providing voltages of 5V, 10V, 12V, 22V and 27V all from a 36V supply. This configuration is not necessarily recommended especially when the current being drawn is seriously mismatched between voltages. It is presented purely out of interest.

## Varactor or Tuning Diodes

These types of diodes work on the principle that all diodes

exhibit some capacitance. Indeed the zener diode BZX79C12 quoted above has, according to the data book, a capacitance of 65-85 pF at 0V and measured at 1 Mhz.

For AM Radio band applications a specific diode has been devised. The Philips BB212 in a TO-92 case is one such type. Each of the diodes has a capacitance of 500 - 620 pF at a reverse bias of 0.5V and <22 pF at 8V. This diode's capacitance ratio is quoted at 22.5:1 which could not be achieved easily if at all with an air variable capacitor.

Several obvious advantages come immediately to mind, a small transistor type package, very low cost, ease of construction on a circuit board, can be mounted away from heat generating devices, frequency determining circuitry entirely dependent upon resistor values and ratios, DC voltage control can be either from frequency synthesiser circuits or perhaps a multi-turn potentiometer. Such a potentiometer aids band spreading and fine tuning if two potentiometers are used. The only real limitation is your imagination and the calculations involved.

## Zener Diodes

The Zener diode is designed to have a specific reverse breakdown voltage (i.e., conduction voltage when reverse-biased). Because of this, Zener diodes can be used by themselves as voltage-sensitive switches, or in series with a current-limiting resistor to provide voltage regulation.

## Photodiodes

All P-N junctions are light sensitive; photodiodes are just P-N junctions that are designed to optimize this effect. Photodiodes can be used two ways -- in a photovoltaic (here it becomes a current source when illuminated — see solar cell), or photoconductive role.

- To use a photodiode in its photoconductive mode, the photodiode is reverse-biased; the photodiode will then allow a current to flow when it is illuminated.

ThermoCentrovision has an interesting site on the technology behind photodiodes here.

## Light-Emitting Diodes (LEDs)

All diodes emit some light when forward-biased. LEDs are made from a special semiconductor (like gallium arsenide phosphide) which optimizes this light output. Unlike light bulbs, LEDs rarely burn out unless their current limit is passed.

When current is flowing through an LED the voltage on the positive leg is about 1.4 volts higher than the voltage on the negative side (this varies with LED type — infrared LEDs have a lower forward voltage requirement, others may need up to 1.8 V). Remember that there is very little resistance to limit the current, so a resistor must be used in series with the LED to avoid destroying it (note, though, that some panel-mount LEDs come from the factory with a current-limiting resistor soldered to them). Also note that LEDs can be used as photodiodes (tho' their sensitivity is relatively low, so they're only useable this way in very bright conditions).

## Flashing LEDs (FLEDs)

A flashing LED is just an LED with a built-in microcircuit to cause it to flash periodically. Since the FLED draws current when it flashes, we can use FLEDs to drive a number of timing-dependent circuits (via the fact that it periodically becomes conductive). In particular, see the discussion of the FLED solarengine design. Like other LEDs, FLEDs are light-sensitive, and so flash faster in brighter light. Note that some

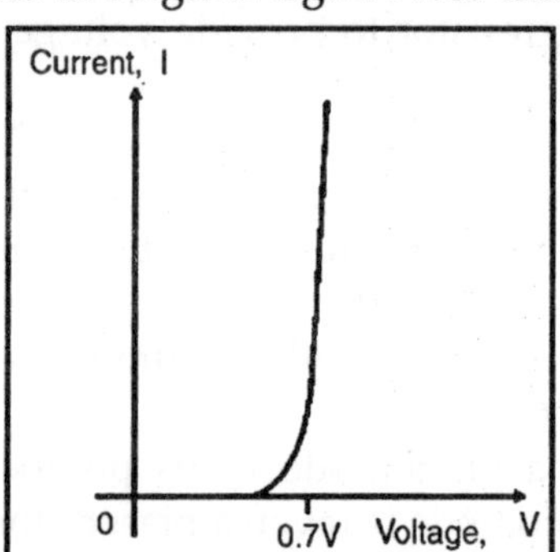


FLEDs need 3 V minimum to work in, but FLEDs don't in general require current-limiting resistors (at least, I've never seen one that does).

### *Function*

Diodes allow electricity to flow in only one direction. The arrow of the circuit symbol shows the direction in which the current can flow. Diodes are the electrical version of a valve and early diodes were actually called valves.

### *Forward Voltage Drop*

Electricity uses up a little energy pushing its way through the diode, rather like a person pushing through a door with a spring. This means that there is a small voltage across a conducting diode, it is called the forward voltage drop and is about 0.7V for all normal diodes which are made from silicon. The forward voltage drop of a diode is almost constant whatever the current passing through the diode so they have a very steep characteristic (current-voltage graph).

### *Reverse Voltage*

When a reverse voltage is applied a perfect diode does not conduct, but all real diodes leak a very tiny current of a few μA or less. This can be ignored in most circuits because it will be very much smaller than the current flowing in the forward direction. However, all diodes have a maximum reverse voltage (usually 50V or more) and if this is exceeded the diode will fail and pass a large current in the reverse direction; this is called breakdown.

Ordinary diodes can be split into two types: Signal diodes which pass small currents of 100mA or less and Rectifier diodes which can pass large currents. In addition there are LEDs (which have their own page) and Zener diodes (at the bottom of this page).

### *Connecting and Soldering*

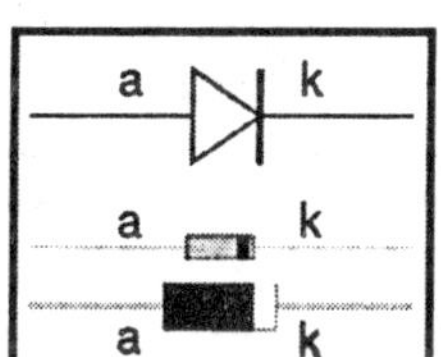

Diodes must be connected the correct way round, the diagram may be labelled a or + for anode and k or - for cathode (yes, it really is k, not c, for cathode!). The cathode is marked by a line painted on the body. Diodes are labelled with their code in small print, you may need a magnifying glass to read this on small signal diodes!

Small signal diodes can be damaged by heat when soldering, but the risk is small unless you are using a germanium diode (codes beginning OA...) in which case you should use a heat sink clipped to the lead between the joint and the diode body. A standard crocodile clip can be used as a heat sink.

Rectifier diodes are quite robust and no special precautions are needed for soldering them.

## Testing Diodes

You can use a multimeter or a simple tester (battery, resistor and LED) to check that a diode conducts in one direction but not the other. A lamp may be used to test a rectifier diode, but do NOT use a lamp to test a signal diode because the large current passed by the lamp will destroy the diode!

## Signal Diodes (Small Current)

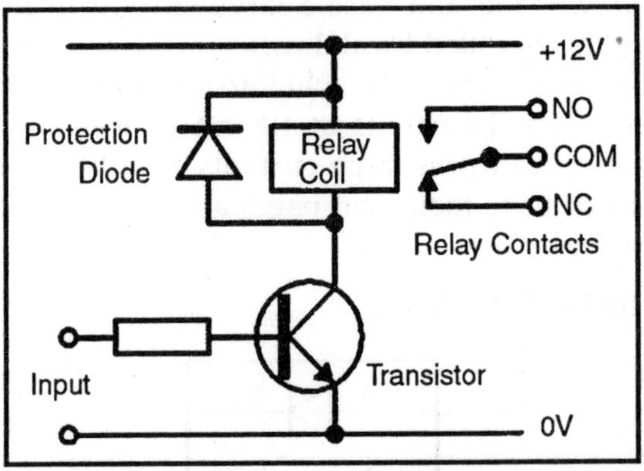


Signal diodes are used to process information (electrical signals) in circuits, so they are only required to pass small currents of up to 100mA. General purpose signal diodes such as the 1N4148 are made from silicon and have a forward voltage drop of 0.7V.

## Germanium Diodes

Such as the OA90 have a lower forward voltage drop of 0.2V and this makes them suitable to use in radio circuits as detectors which extract the audio signal from the weak radio signal.

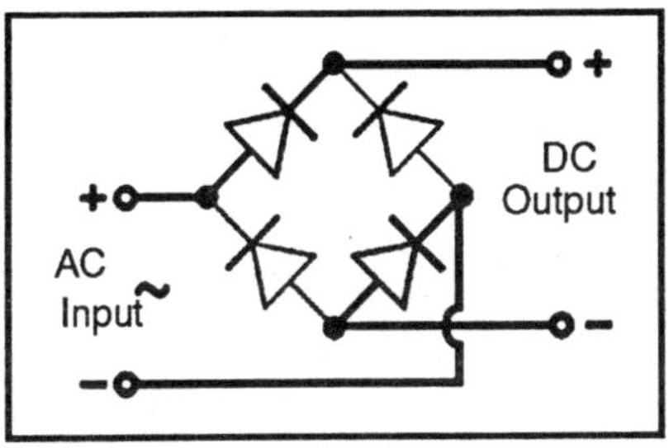


For general use, where the size of the forward voltage drop is less important, silicon diodes are better because they are less easily damaged by heat when soldering, they have a lower resistance when conducting, and they have very low leakage currents when a reverse voltage is applied.

## Protection Diodes

Signal diodes are also used to protect transistors and ICs from the brief high voltage produced when a relay coil is switched off. The diagram shows how a protection diode is connected 'backwards' across the relay coil.

Current flowing through a relay coil creates a magnetic field which collapses suddenly when the current is switched off. The sudden collapse of the magnetic field induces a brief high voltage across the relay coil which is very likely to damage transistors and ICs.

The protection diode allows the induced voltage to drive a brief current through the coil (and diode) so the magnetic field dies away quickly rather than instantly. This prevents the induced voltage becoming high enough to cause damage to transistors and ICs.

| Diode | Maximum Current | Maximum Reverse Voltage |
|---|---|---|
| 1N4001 | 1A | 50V |
| 1N4002 | 1A | 100V |
| 1N4007 | 1A | 1000V |
| 1N5401 | 3A | 100V |
| 1N5408 | 3A | 1000V |

## Rectifier Diodes

Rectifier diodes are used in power supplies to convert alternating current (AC) to direct current (DC), a process called rectification. They are also used elsewhere in circuits where a large current must pass through the diode.

All rectifier diodes are made from silicon and therefore have a forward voltage drop of 0.7V. The table shows maximum current and maximum reverse voltage for some popular rectifier diodes. The 1N4001 is suitable for most low voltage circuits with a current of less than 1A.

## DIODES AS FREQUENCY MULTIPLIERS

Just one more example of the versatility of diodes is the frquency doubling circuit depicted in Figure below. Now if that looks a lot like the full wave rectifier, it would be correct. That is why the ripple frequency for 50 / 60 Hz always comes out at 100 / 120 Hz.

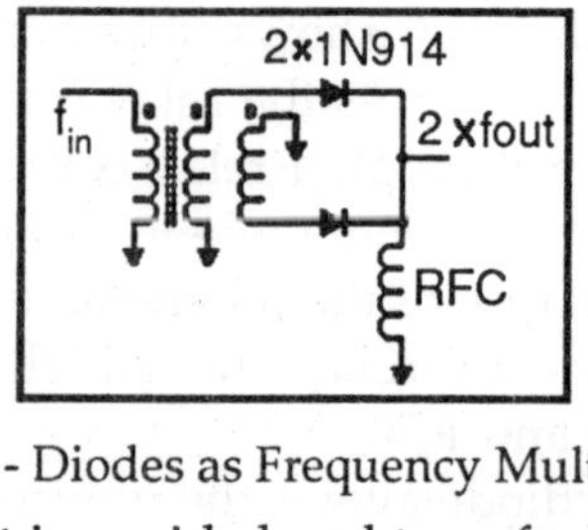


Figure - Diodes as Frequency Multipliers

Here the input is a wide band transformer and the signal passes to a full wave rectifier comprising two 1N914 diodes. The DC component caused by the rectification passes to ground through RFC which of course presents a high impedance to the rf porion of the signal but essentially a short circuit for DC.

The original signal should be down about 40 dB and with this type of circuit there would be a loss of somewhere around 7.5 dB so the 2 X signal would require further amplification to restore that loss.

## ZENER DIODES

*Example:*

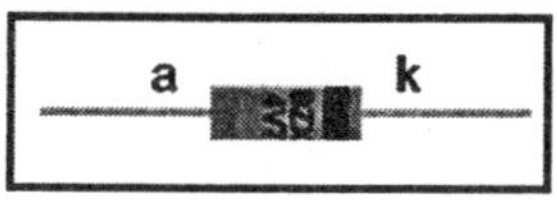


*Circuit Symbol:*

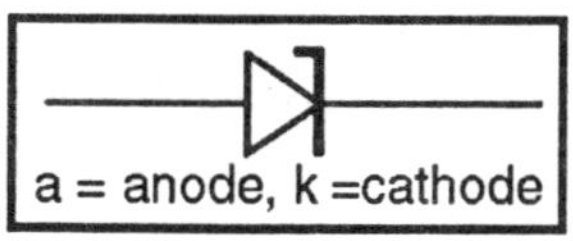


a = anode, k = cathode

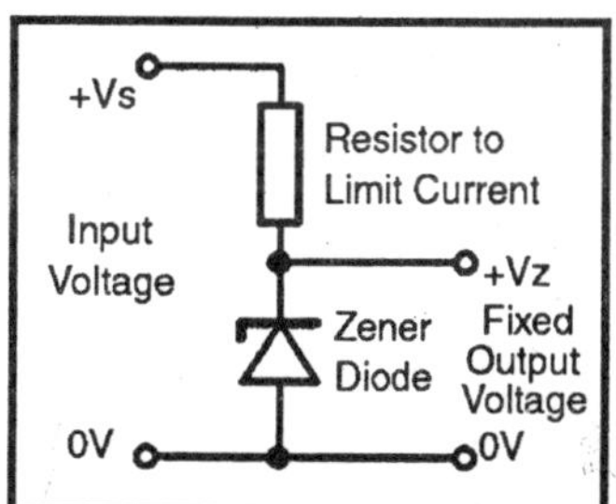


Zener diodes are used to maintain a fixed voltage. They are designed to 'breakdown' in a reliable and non-destructive way so that they can be used in reverse to maintain a fixed voltage across their terminals. The diagram shows how they are connected, with a resistor in series to limit the current.

Zener diodes can be distinguished from ordinary diodes by their code and breakdown voltage which are printed on them. Zener diode codes begin BZX... or BZY... Their breakdown voltage is printed with V in place of a decimal point, so 4V7 means 4.7V for example.

Zener diodes are rated by their breakdown voltage and maximum power:

- The minimum voltage available is 2.4V.
- Power ratings of 400mW and 1.3W are common.

## DESCRIPTION

Diodes are semiconductor devices which might be described as passing current in one direction only. The latter part of that statement applies equally to vacuum tube diodes. Diodes however are far more versatile devices than that. They are extremely versatile in fact

Diodes can be used as voltage regulators, tuning devices in rf tuned circuits, frequency multiplying devices in rf circuits, mixing devices in rf circuits, switching applications or can be used to make logic decisions in digital circuits. There are also diodes which emit "light", of course these are known as light-emitting-diodes or LED's. As we say diodes are extremely versatile.

## SCHEMATIC SYMBOLS FOR DIODES

*A few schematic symbols for diodes are:*

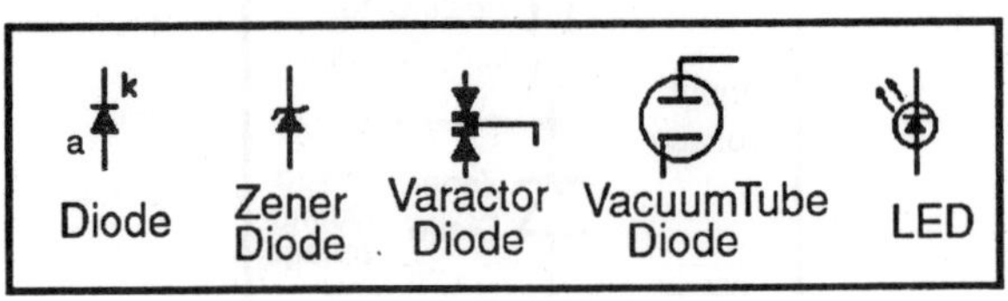


**Fig.** Schematic Symbols for Diodes

## APPLICATIONS OF DIODES

Similar types of diodes have been developed specifically for band switching purposes. Although a typical 1N914 type switching diode can be used for such purposes it is preferable to use diodes which have been optimised for such purposes because the Rd on is much lower.

This means the diode resistance Rd can have a serious affect on rf circuits in particular the "Q" of a tuned circuit. One example of a low Rd device is the Philips BA482 diode used for band switching in television tuners. It has a typical Rd of 0.4 ohms at a forward current of 10 mA.

In figure 5 we have one application where switching diodes operate. All diodes serve to switch in or out capacitors in the diagram which is presented here just to illustrate one single application of switching diodes, many, many more applications exist. Again the limit is your imagination.

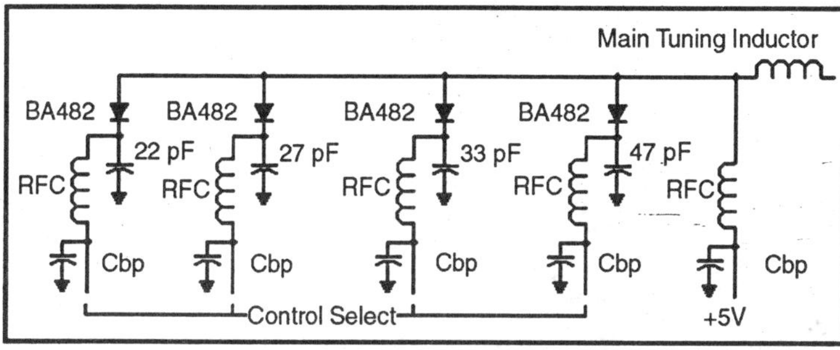


**Fig.** Applications of Switching Diodes.

The switching diodes in figure 4 switch in or out successively higher values of capacitors as each control select line is "grounded".

The voltage from the +5V feed line proceeds through the diode at DC thus opening the diode and making it appear "transparent" for rf purposes.

The capacitor with the value attached is then "switched" into circuit. Other components marked RFC and Cbp are chokes and bypass capacitors for "clean" switching. The bypass capacitors and choke values would be determined by the frequency of operation.

We could just have easily have switched inductors instead of capacitors. Note why Rd is quite important on overall circuit performance.

If we were using inductors the diode resistance Rd would have a significant affect on inductor "Q" which in turn would affect filter performance, if it was in fact an LC filter application.

## BUILDING THE H-BRIDGE

The complete schematic for the bridge is shown below.

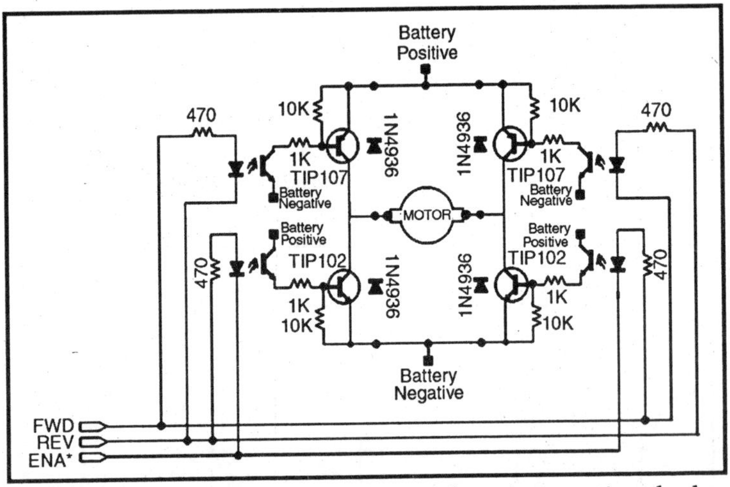


The fast recovery diodes (shown above) are optional when you are using the TIP102/107 as those transistors have a diode built into them. However you can include them for greater protection at higher currents. You will note that I also take advantage of the symmetrical nature of the bridge by using resistor packs rather than individual resistors.

## Processor Interface

The FWD, REV, and ENA* lines are the interface between the bridge and the microprocessor. You will notice their is no "ground" signal. When you connect these pins to a BASIC stamp or a 68HC11 or what-ever, combinations of 1's and 0's on the line turn on different pairs of transistors. The following table lists all possible combinations of input.

| FWD | REV | ENA* | Description |
|---|---|---|---|
| 1 | 0 | 0 | Turn on upper left source and lower right sink. (go forward) |
| 1 | 0 | 1 | Disable lower right sink. When fed a PWM signal the bridge modulates the "forward" current through the motor. |
| 1 | 1 | 0 | Turn on both lower left sink |

| | | | |
|---|---|---|---|
| | | | and lower right sink, shorting the motor. This causes a rotating motor to stop rotating so this mode is called "Braking." |
| 1 | 1 | 1 | Disable both lower sinks. When fed a PWM signal the bridge modulates the "braking" of the motor. |
| 0 | 1 | 0 | Turn on the upper right source and lower left sink. (go backward) |
| 0 | 1 | 1 | Disable lower left sink. When fed a PWM signal the bridge modulates the "reverse" current through the motor. |
| 0 | 0 | 0 | Turn off all sources and sinks. "Coast" motor is not engaged at all. |
| 0 | 0 | 1 | Turn off all sources and sinks in a different way, same effect though. |

This works because the processor pins become a connection to ground when they are outputting a logic 0. Thus when FWD is 1 and ENA is 0, the lower right sink is getting current from FWD which it is returning through the pin connected to ENA. At any given time the pin must be able to supply enough current to turn on two LEDs and when set to zero sink the current of two LEDs. With the 560 ohm resistors and a 5V processor like the PIC this is not a problem.

From the data sheet on the opto-isolator, each LED has a forward voltage drop of 1.2V, so (5 - 1.2) / 560 is 6.8mA per LED or 14mA of load total. The PIC is specified to be able to drive 20mA.

You can replace the 560 ohm resistors with 680 ohm resistors to reduce the current through the LEDs still further. However the LED voltage drop can be as high as 1.4V and the PIC's operating voltage can be as low as 4.5V, in that condition with 680 ohm resistors you would only put 4.5mA into each LED which is below the 5mA specified in the datasheet. The 560 ohm resistor is the better choice as it maintains sufficient margin. So why go to all this trouble? The interface as designed gives you access to all of the interesting combinations of sinks and sources enable, while not allowing for any "illegal" states. An illegal state in a full quadrant H-bridge (4 inputs, one for each quadrant) is one that turns on the upper source and lower sink on the same side.

This combination causes a direct short circuit to be created between the battery terminals (not good!) and usually causes one or both of the transistors that are on to go up in smoke. Further, as the low end PIC chips like the PIC16F628 has a single hardware PWM line, this lets you connect that line to the bridge and a couple of GPIO pins to the FWD and REV lines to get a hardware PWM on the bridge. Just remember not to go above 2.5Khz unless you replace the opto-isolator with a faster one.

## CHARACTERISTICS

### FAN-IN

Fan-in is number of inputs a gate has, Like a two input AND gate has fan-in of two, a three input NAND gate as a fan-in of three. So a NOT gate always has a fan-in of one. Below figure shows the effect of Fan-in on the delay offered by gate for a CMOS based gate. Normally delay increases quadratic function of fan-in.

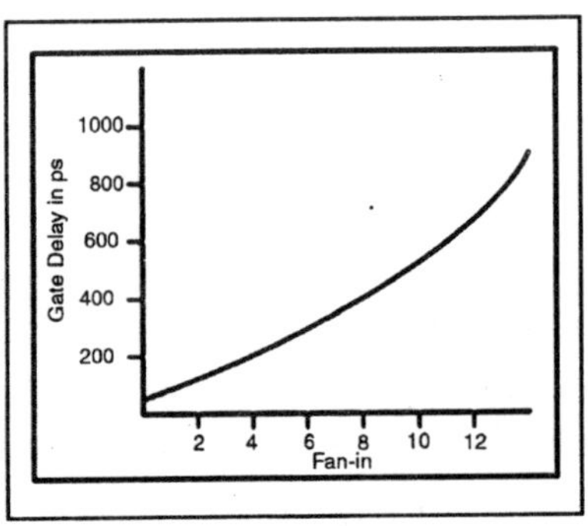

## FAN-OUT

The number of gates that each gate can drive, while providing voltage levels in the guaranteed range is called the standard load or fan-out. The fan-out really depends on the amount of electric current a gate can source or sink while driving other gates. The effects of loading a logic gate output with more than its rated fan-out has the following effects.

- In the LOW state the output voltage VOL may increase above VOLmax.
- In the HIGH state the output voltage VOH may decrease below VOHmin.
- The operating temperature of the device may increase thereby reducing the reliability of the device and eventually causing the device failure.
- Output rise and fall times may increase beyond specifications

The propagation delay may rise above the specified value.

Normally as in the case of fan-in, delay offered by a gate increases with increase in fan-out.

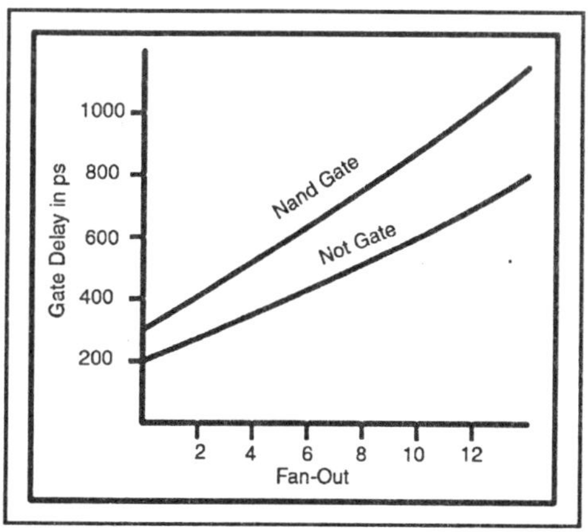


## TRI-STATE LOGIC

## INTRODUCTION

Both combinatorial circuits and sequential circuits are

fairly easy to understand, since we pretty much only have to understand logic gates and how they are interconnected.

With tri-state logic circuits, this is no longer true. As their names indicate, they manipulate signals that can be in one of three states, as opposed to only 0 or 1. While this may sound confusing at first, the idea is relatively simple.

Consider a fairly common case in which there are a number of source circuits $S_1$, $S_2$, etc in different parts of a chip (i.e., they are not real close together).

At different times, exactly one of these circuit will generate some binary value that is to be distributed to some set of destination circuits $D_1$, $D_2$, etc also in different parts of the chip.

At any point in time, exactly one source circuit can generate a value, and the value is always to be distributed to all the destination circuits.

Obviously, we have to have some signals that select which source circuit is to generate information. Assume for the moment that we have singals $s_1$, $s_2$, etc for exactly that purpose. One solution to this problem is indicated in this figure:

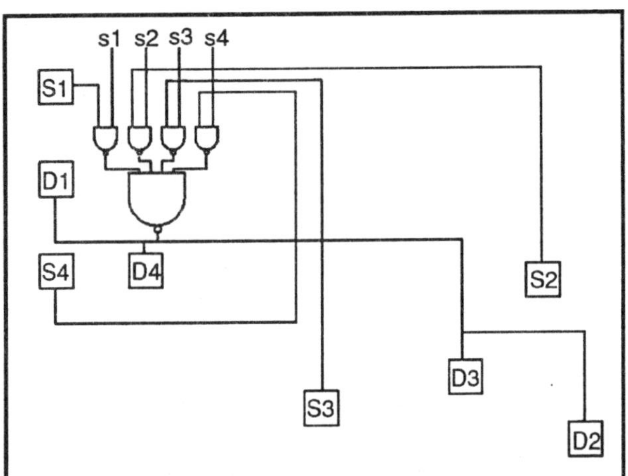


As you can see, this solution requires that all outputs are routed to a central place. Often such solutions are impractical or costly. Since only one of the sources is "active" at one point, we ought to be able to use a solution like this:

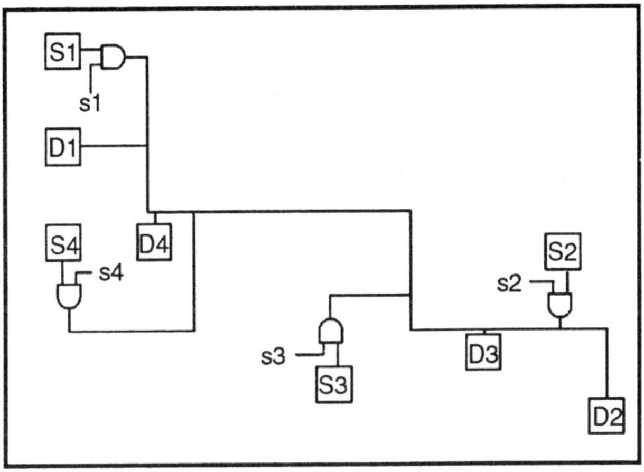


However, connecting two or more outputs together is likely to destroy the circuits. The solution to our problem is to use tri-state logic.

## TRANSISTORS

A tri-state circuit (combinatorial or sequential) is like an ordinary circuit, except that it has an additional input that we shall call enable. When the enable input is 1, the circuit behaves exactly like the corresponding normal (not tri-state) circuit. When the enable input is 0, the outputs are completely disconnected from the rest of the circuit. It is as if there we had taken an ordinary circuit and added a switch on every output, such that the switch is open when enable is 0 and closed if enable is 1 like this:

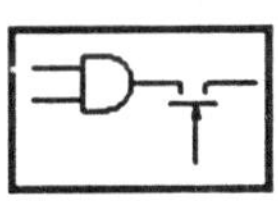

which is pretty close to the truth. The switch is just another transistor that can be added at a very small cost.

Any circuit can exist in a tri-state version. However, as a special case, we can convert any ordinary circuit to a tri-state circuit, by using a special tri-state combinatorial circuit that simply copies its inputs to the outputs, but that also has an enable input. We call such a circuit a bus driver for reasons that will become evident when we discuss buses. A bus driver with one input is drawn like this:

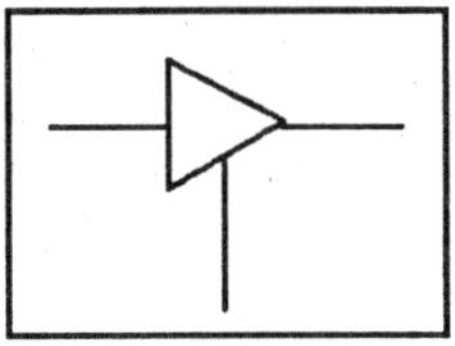

Here is a version for bidirectional signals:

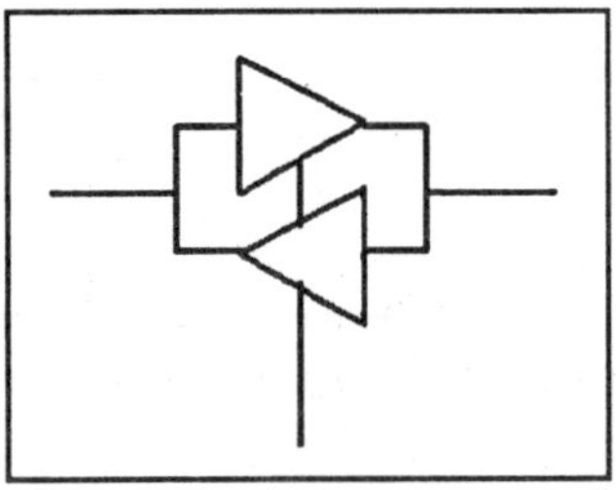

## GATE DELAYS

As an example of the level of sophistication of Logisim's algorithm, consider the following circuit.

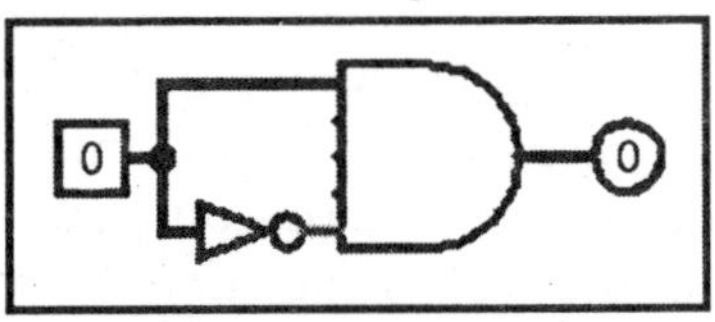


This "obviously" always outputs 0. But NOT gates don't react instantaneously to their inputs in reality, and neither do they in Logisim. As a result, when this circuit's input changes from 0 to 1, the AND gate will briefly see two 1 inputs, and it will emit a 1 briefly. You won't see it on the screen. But the effect is observable when we use the AND gate's output as an input into the clock of a D flip-flop.

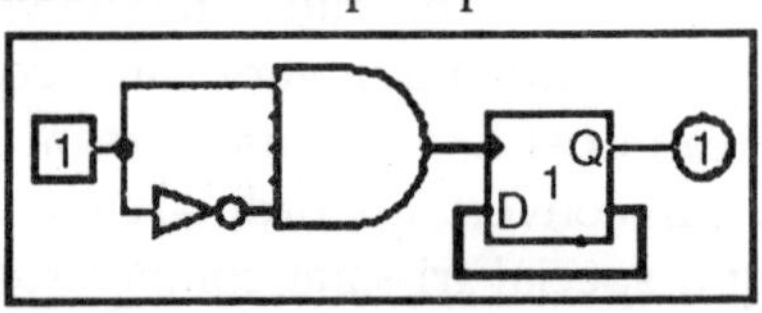


Poking the 0 input to become 1 leads to an instantaneous 1 going into the D flip-flop, and thus the flip-flop's value will toggle every time the circuit input goes from 0 to 1. Every component has a delay associated with it. More sophisticated components built into Logisim tend to have larger delays, but these delays are somewhat arbitrary and may not reflect reality.

# GENERATION OF SPIKES

## INTRODUCTION

Each living creature has to act in a suitable way in response to the various situations it encounters in its environment. Nowadays it is accepted that the nervous system (including the spinal cord as well as the neocortex) controls the behaviour.

In a simplified picture one could view the nervous system as a device which receives input from various senses like auditory input from the ears and produces (computes) some output in the form of movement or speech.

To understand how this computations are performed is one of the most interesting and challenging tasks in science. Thus it is not surprising that the field of neuroscience attracts a lot of researchers not only from biology but also from physics and computer science.

In this article we try to give the reader an intuition what computer scientists (especially at the Institute for Theoretical Computer Science at the Technical University of Graz) can contribute to the question how the brain works.

The central questions of a theoretical computer scientist are: How much (computational) power is needed to compute a certain function? Which functions can be computed by a given „device"? In the context of this article a concrete question could be: How many nerve cells are needed to be able to distinguish between a picture of your mother and your father?

However, the questions addressed in this article are still on a more abstract level since theoretical computer scientists try to make general statements about computation.

An article written by W. Maass investigates this kind of questions in a more mathematical manner.

## BIOLOGICAL NEURAL SYSTEMS

One of the most important parts of a neuron is its membrane, which separates the interior and the exterior of a neuron spatially and electrically.

A prominent feature of the cell membrane is, that it maintains a voltage difference between the interior and the exterior.

This so called membrane potential is subject to change if a neuron receives some input via its synapses (manly located at the dendritic tree of a neuron) from other neurons.

There are two basic types of input: excitatory input (via excitatory synapses) increases the membrane potential whereas inhibitory input (via inhibitory synapses) decreases the membrane potential.

If enough excitatory input accumulates the membrane potential at the axon hillock eventually reaches a threshold. If this event occurs the membrane follows a stereotyped trajectory, the so called action potential or spike.

This spike travels along the axon (the "output wire" of a neuron) to all its terminals which connects the axon to other neurons (postsynaptic neurons) via synapses.

A complicated chemical process transforms the spike into a change of the membrane potential of the postsynaptic neuron, the so called postsynaptic potential (PSP) which comprise the above mentioned input of the next neuron.

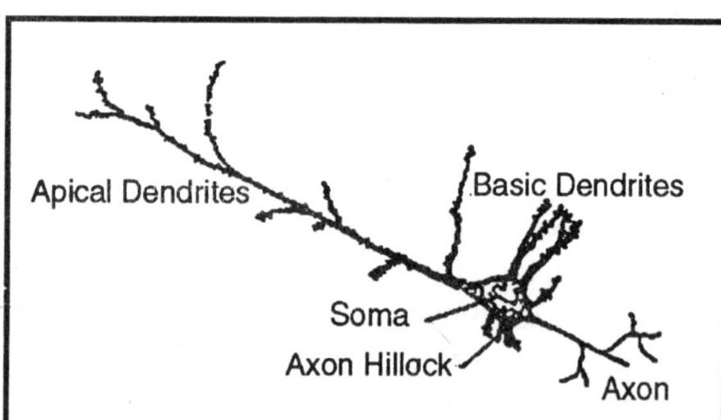


**Fig.** Schematic picture of a nerve cell

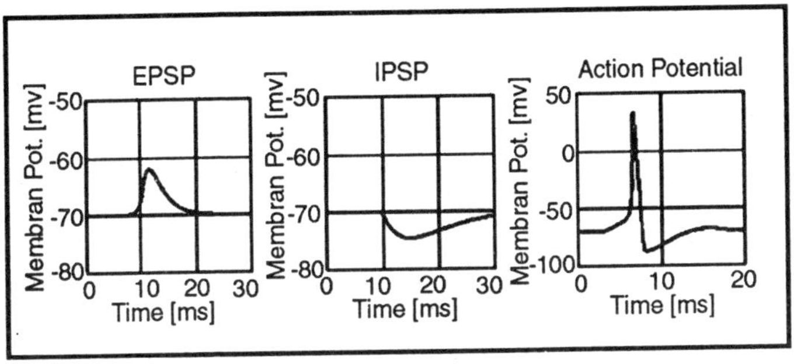


**Fig.** Drawing of the Canonical Shape of a Postsynaptic Potential and the Action Potential.

## First Generation of Models

If one wants to understand how the nervous system computes a function one has to think about how information about the environment or internal states is represented and transmitted. The fact that the shape of an action potential is always the same one can exclude the possibility that the voltage trajectory of an action potential carries relevant information. Thus a central question in the field of neuroscience is how neurons encode information in the sequence of action potential they emit.

1943 McCulloch and Pitts proposed the first neuron model: the threshold gate. The characteristics of their model was that they treated a neuron as a binary device. That is they distinguished only between the occurrence and absence of a spike. The threshold gate is used as building block for various network types including multilayer perceptrons, Hopfield networks and the Boltzman machine. It turned out that the threshold gate is a computational powerful device. That is one can compute complex functions with rather small networks made up of threshold gates. From the theoretical point of view the threshold gate is a very interesting model but it is unlikely that real biological systems use such a binary encoding scheme. A prerequisite for such a binary coding scheme is a kind of global clocking mechanism but it is very unlikely that such a mechanism exists in biological systems.

## Second Generation

Another possibility is that the number of spikes per second (called the firing rate) encodes relevant information. This idea lead to a model neuron known as sigmoidal gate. The output of a sigmoidal gate is a number which is thought to represent the firing rate of the neuron.

There exists a huge amount of literature which discusses in detail all aspects of this kind of neural network models. We just want to note that networks of sigmoidal gates can in principle compute any analog function and that along with this type of models the question of learning in neural networks was intensively investigated for the first time.

### *Firing Rates versus Spike Timing*

It is well known that firing rates play an important role in the nervous system especially in the primary sensory areas where low level information processing takes place. However, recent experiments revealed that the human nervous system is able to perform complex visual tasks in time intervals as short as 150–ms (1–ms = 0.001 second). The pathway from the retina to higher areas in the neocortex along which a visual stimuli is processed consists of about 10 "processing stages".

Furthermore the firing rates in the areas involved in such complex computations are well below 100 spikes per second. But to estimate such low firing rates one has to wait at least 30 to 50–ms which is a contradiction to the finding that a computation involving 10 "processing stages" can be carried out within 150°ms.

## Third Generation: Networks of Spiking Neurons

These results from experimental neurobiology gave rise to a new class of neural network models where one also incorporates the timing of individual spikes. Thus time plays a central role in SNNs whereas in most other neural network models there is even no notion of time.

### *Temporal Coding*

The inputs to a network of spiking neurons are encoded

in the time differences between a spike and a reference time point. The output is also encoded by the time difference between an output spike and some other reference time point. The reference time points can be given by other spikes but also through other mechanisms, e.g. background oscillations.

## *Computational Power*

What class of functions can be computed by a SNN using this type of temporal coding? An important result regarding this question is reported by Maass, 1997. In this work it is shown that a small network of spiking neurons using temporal coding can easily compute the same functions as sigmoidal gate.

As noted above a network of sigmoidal function can in principle compute any given function and hence also a network of spiking neurons can compute any given function regardless how complex this function might be. An interesting detail of this result is that the time which the SNN needs to perform such a computation is in the order of 100 ms.

At least from a theoretical point of view this result shows a possibility how complex computations in the human nervous system might be organized. In another article it shown that SNNs have even more computational power than networks of sigmoidal gates. That means that there are function which can easily be computed by a small network of spiking neurons but one needs a large net of sigmoidal gates.

## *Hypothesis for Biological Neural Systems*

In the previous subsection we pointed out that SNNs have enough computational power to compute arbitrary functions. However the proofs underlying these results are done on a rather abstract level and furthermore do not provide hints how to construct a network of spiking neurons which computes a concrete function. But on the other hand if it would had turned out that an SNN using temporal coding has very low computational power this would rule out such a temporal code as potential mechanism used in real biological systems.

The question about the computational power is not the

only one which is investigated at the Institute for Theoretical Computer Science. We are also interested how concrete function which are likely to be performed in real biological neural systems could be implemented efficiently in a network of spiking neurons. An example of such a function is the associative recall of stored memories.

In the work of Maass and Natschläger, 1997 it is show how such a associative memory can be implemented with a biological rather realistic network of spiking neurons.

Another example of a concrete problem is given by Hopfield, 1995. In his work he comes up with a hypothesis how a simple form of pattern recognition might be implemented with a network of spiking neurons. A good survey about models for concrete functions or subsystems of the nervous systems can be found in the textbook by Koch and Segev, 1998.

Such hypothesis about the functionality of certain subsystems of the nervous system may led to new experiments the results of which allow one to make more realistic models. Thus the interplay between the experiments done by biologists and theoretical models made by physicists or computer scientists is a very fruitful interdisciplinary interplay.

### *Learning*

One of the most challenging problems in neuroscience is the question of learning. Although substantial progress has been made regarding learning in neural networks of the second generation these "learning schemes" do not directly apply to biological systems (see the textbook by Bishop for an overview). What is commonly accepted is that learning takes place in the connections between to neurons, i.e. the synapses are the place where changes are made if we learn something.

Until recently it was assumed that the changes in a synapse depend basically on the relation between the firing rates of the two neurons a synapse connects. However results reported by Markram et. al. indicate that the changes in a synapse depend strongly on the timing of the individual spikes the two neurons produce. Based on this experimental results several models have been developed how such a time depended learning scheme my be used in a network of spiking neurons to learn something meaningful.

An example is the work of Natschläger and Ruf, 1997 where it is shown that such a learning scheme might be useful to extract important features from a stimulus. Although there exist many other hypothesis regarding such a time depended learning scheme non of these hypothesis can explain high level memory phenomena.

This is a characteristic phenomena in the whole field of neuroscience. One knows a lot about biological neural systems from the molecular to the cellular level, but we have basically no idea how these bits and pieces together form the most powerful computational device which we are aware of.

### *Applications*

Network models of the second generation turned out to be a very useful tool for a wide variety of problems. This gave rise to the development of fast hardware implementations of such models to be able to solve large problems in a reasonable time.

One approach is to use analogue devices instead of microprocessors to carry out the computations. It turned out to be advantageous if such "silicon neurons" communicate via voltage pulses.

An example of such a hardware implementation is EPSILON II, a chip developed at the University of Edinburgh. This chip does not only use pulses for communication purposes but also employees the timing of individual pulses for computation.

This example shows that the principles underlying the information processing in biological neural systems also inspires engineers when designed new hardware. In the last view years people all over the world started to develop such "neuro-inspired" (also called neuromorphic) systems. The idea behind this efforts is to reveal solutions for problems that can not be solved yet efficiently but it is evident that mother nature has developed very efficient solutions for.

# Chapter 7

# Memories

## STATIC AND DYNAMIC RAM

### STATIC RAM (SRAM)

Static RAM is a type of RAM that holds its data without external refresh, for as long as power is supplied to the circuit. This is contrasted to dynamic RAM (DRAM), which must be refreshed many times per second in order to hold its data contents. SRAMs are used for specific applications within the PC, where their strengths outweigh their weaknesses compared to DRAM:

- *Simplicity*: SRAMs don't require external refresh circuitry or other work in order for them to keep their data intact.
- *Speed*: SRAM is faster than DRAM.

In contrast, SRAMs have the following weaknesses, compared to DRAMs:

- *Cost*: SRAM is, byte for byte, several times more expensive than DRAM.
- *Size*: SRAMs take up much more space than DRAMs (which is part of why the cost is higher).

These advantages and disadvantages taken together obviously show that performance-wise, SRAM is superior to DRAM, and we would use it exclusively if only we could do so economically.

Unfortunately, 32 MB of SRAM would be prohibitively large and costly, which is why DRAM is used for system memory. SRAMs are used instead for level 1 cache and level 2

Of course, totem-pole output stages are also possible in both NOR and OR TTL logic circuits.

*Review*:

- An OR gate may be created by adding an inverter stage to the output of the NOR gate circuit.

## CMOS GATE CIRCUITRY

Up until this point, our analysis of transistor logic circuits has been limited to the TTL design paradigm, whereby bipolar transistors are used, and the general strategy of floating inputs being equivalent to "high" (connected to $V_{cc}$) inputs — and correspondingly, the allowance of "open-collector" output stages — is maintained. This, however, is not the only way we can build logic gates. Field-effect transistors, particularly the insulated-gate variety, may be used in the design of gate circuits. Being voltage-controlled rather than current-controlled devices, IGFETs tend to allow very simple circuit designs. Take for instance, the following inverter circuit built using P- and N-channel IGFETs:

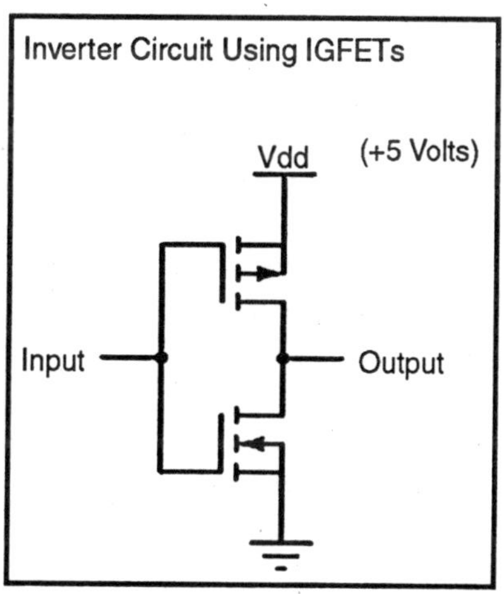


Notice the "$V_{dd}$" label on the positive power supply terminal. This label follows the same convention as "$V_{cc}$" in TTL circuits: it stands for the constant voltage applied to the drain of a field effect transistor, in reference to ground.

Let's connect this gate circuit to a power source and input

switch, and examine its operation. Please note that these IGFET transistors are E-type (Enhancement-mode), and so are normally-off devices. It takes an applied voltage between gate and drain (actually, between gate and substrate) of the correct polarity to bias them on.

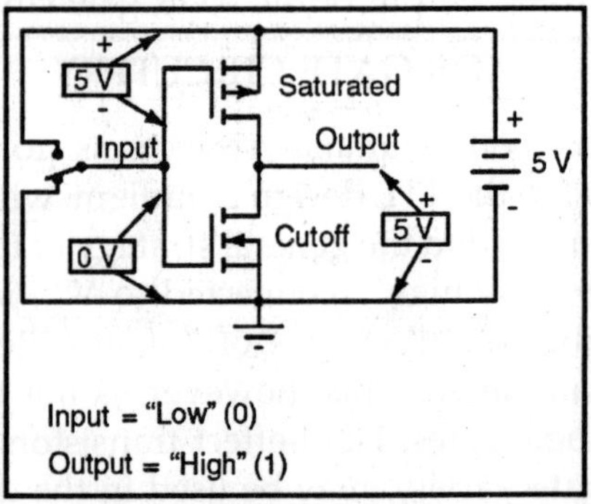


The upper transistor is a P-channel IGFET. When the channel (substrate) is made more positive than the gate (gate negative in reference to the substrate), the channel is enhanced and current is allowed between source and drain. So, in the above illustration, the top transistor is turned on.

The lower transistor, having zero voltage between gate and substrate (source), is in its normal mode: off. Thus, the action of these two transistors are such that the output terminal of the gate circuit has a solid connection to $V_{dd}$ and a very high resistance connection to ground. This makes the output "high" (1) for the "low" (0) state of the input. Next, we'll move the input switch to its other position and see what happens:

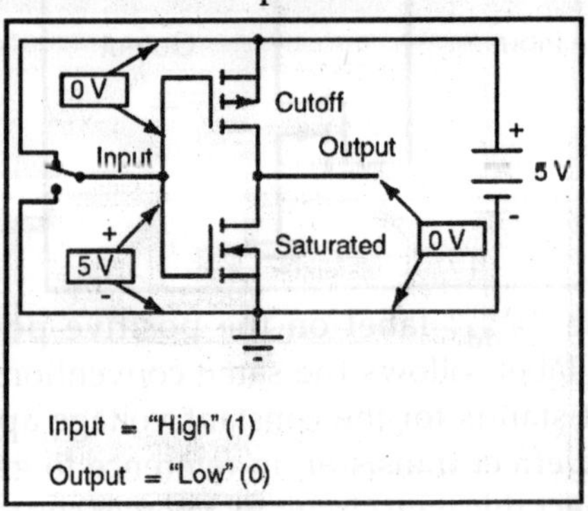

Now the lower transistor (N-channel) is saturated because it has sufficient voltage of the correct polarity applied between gate and substrate (channel) to turn it on (positive on gate, negative on the channel).

]The upper transistor, having zero voltage applied between its gate and substrate, is in its normal mode: off. Thus, the output of this gate circuit is now "low" (0). Clearly, this circuit exhibits the behaviour of an inverter, or NOT gate.

Using field-effect transistors instead of bipolar transistors has greatly simplified the design of the inverter gate. Note that the output of this gate never floats as is the case with the simplest TTL circuit: it has a natural "totem-pole" configuration, capable of both sourcing and sinking load current.

Key to this gate circuit's elegant design is the complementary use of both P- and N-channel IGFETs. Since IGFETs are more commonly known as MOSFETs (Metal-Oxide-Semiconductor Field Effect Transistor), and this circuit uses both P- and N-channel transistors together, the general classification given to gate circuits like this one is CMOS: Complementary Metal Oxide Semiconductor.

CMOS circuits aren't plagued by the inherent nonlinearities of the field-effect transistors, because as digital circuits their transistors always operate in either the saturated or cutoff modes and never in the active mode.

Their inputs are, however, sensitive to high voltages generated by electrostatic (static electricity) sources, and may even be activated into "high" (1) or "low" (0) states by spurious voltage sources if left floating.

For this reason, it is inadvisable to allow a CMOS logic gate input to float under any circumstances. Please note that this is very different from the behaviour of a TTL gate where a floating input was safely interpreted as a "high" (1) logic level.

This may cause a problem if the input to a CMOS logic gate is driven by a single-throw switch, where one state has the input solidly connected to either $V_{dd}$ or ground and the other state has the input floating (not connected to anything):

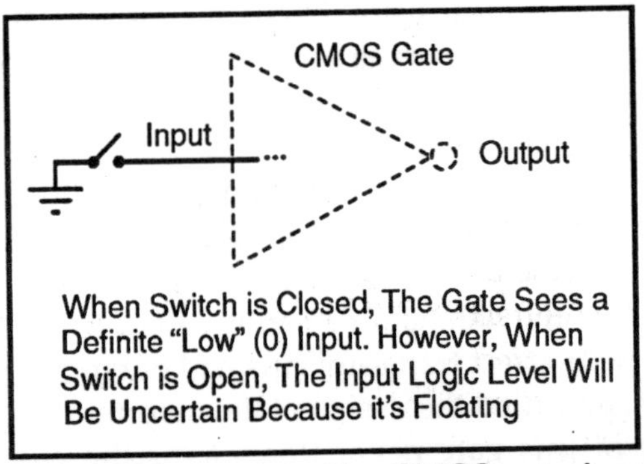


Also, this problem arises if a CMOS gate input is being driven by an open-collector TTL gate. Because such a TTL gate's output floats when it goes "high" (1), the CMOS gate input will be left in an uncertain state:

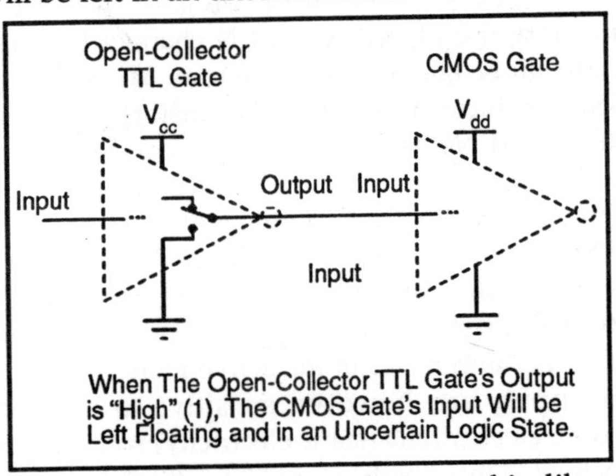


Fortunately, there is an easy solution to this dilemma, one that is used frequently in CMOS logic circuitry. Whenever a single-throw switch (or any other sort of gate output incapable of both sourcing and sinking current) is being used to drive a CMOS input, a resistor connected to either $V_{dd}$ or ground may be used to provide a stable logic level for the state in which the driving device's output is floating.

This resistor's value is not critical: 10 kÙ is usually sufficient. When used to provide a "high" (1) logic level in the event of a floating signal source, this resistor is known as a pullup resistor:

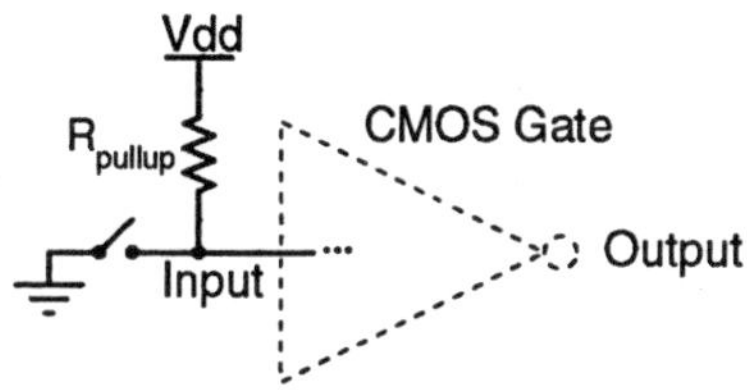


When Switch is closed, the Gate sees a Definite "Low" (0) input.When the switch is open, $R_{pullup}$ will provide the Connection to Vdd needed to secure a Reliable "High" logic level for the CMOS Gate input.

When such a resistor is used to provide a "low" (0) logic level in the event of a floating signal source, it is known as a pulldown resistor. Again, the value for a pulldown resistor is not critical:

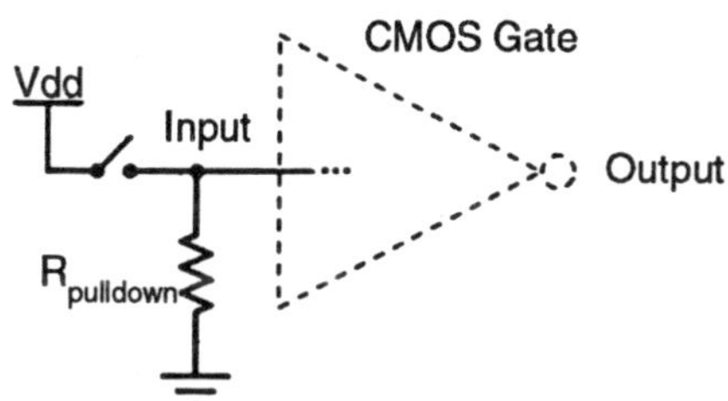


When Switch is closed,the Gate sees a Definite "High"(1)input. When the switch is open,$R_{pulldown}$ will provide the Connection to ground needed to secure a Reliable "Low" logic level for the CMOS Gate input

Because open-collector TTL outputs always sink, never source, current, pullup resistors are necessary when interfacing such an output to a CMOS gate input:

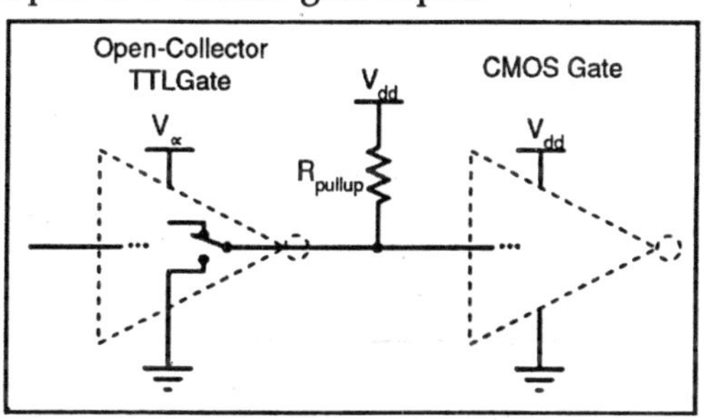

Although the CMOS gates used in the preceding examples were all inverters (single-input), the same principle of pullup and pulldown resistors applies to multiple-input CMOS gates. Of course, a separate pullup or pulldown resistor will be required for each gate input:

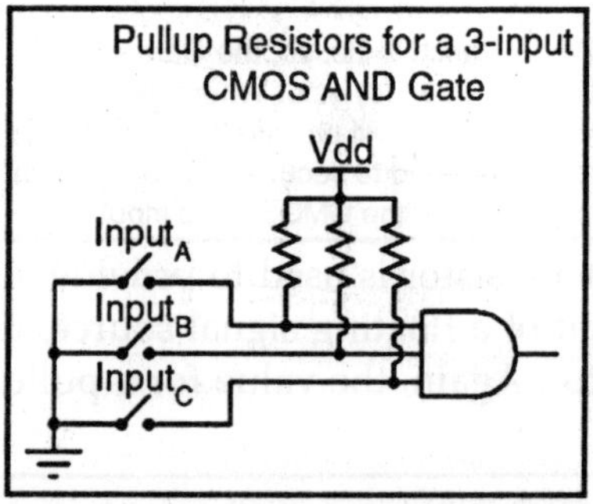


This brings us to the next question: how do we design multiple-input CMOS gates such as AND, NAND, OR, and NOR? Not surprisingly, the answer(s) to this question reveal a simplicity of design much like that of the CMOS inverter over its TTL equivalent. For example, here is the schematic diagram for a CMOS NAND gate:

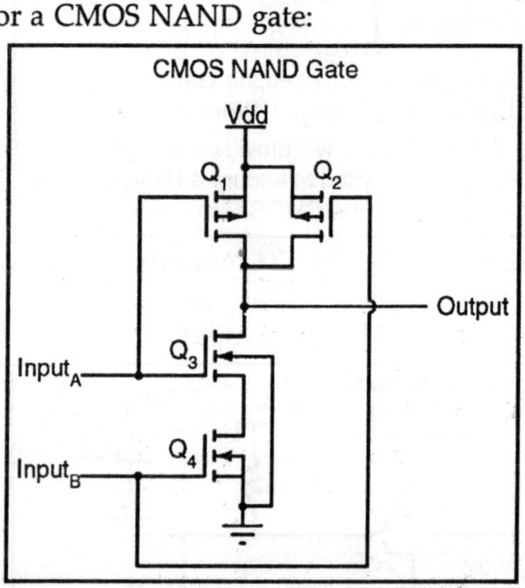


Notice how transistors $Q_1$ and $Q_3$ resemble the series-connected complementary pair from the inverter circuit. Both are controlled by the same input signal (input A), the upper transistor turning off and the lower transistor turning on when the input is "high" (1), and visa-versa. Notice also how transistors $Q_2$ and $Q_4$ are similarly controlled by the same input signal (input B), and how they will also exhibit the same on/off behaviour for the same input logic levels. The upper transistors of both pairs ($Q_1$ and $Q_2$) have their source and drain terminals paralleled, while the lower transistors ($Q_3$ and $Q_4$) are series-connected. What this means is that the output will go "high" (1) if either top transistor saturates, and will go "low" (0) only if both lower transistors saturate. The following sequence of illustrations shows the behaviour of this NAND gate for all four possibilities of input logic levels (00, 01, 10, and 11):

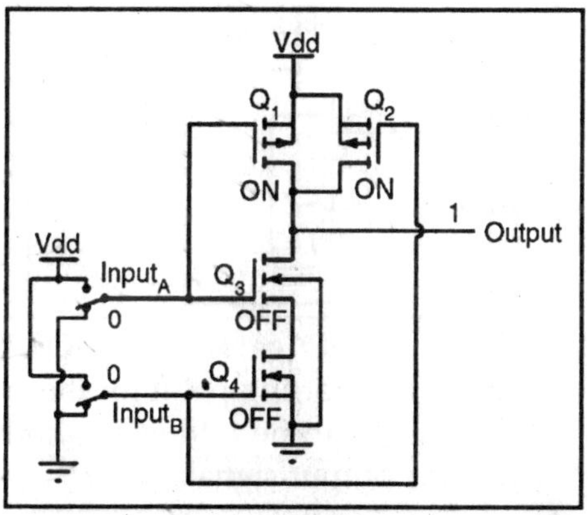


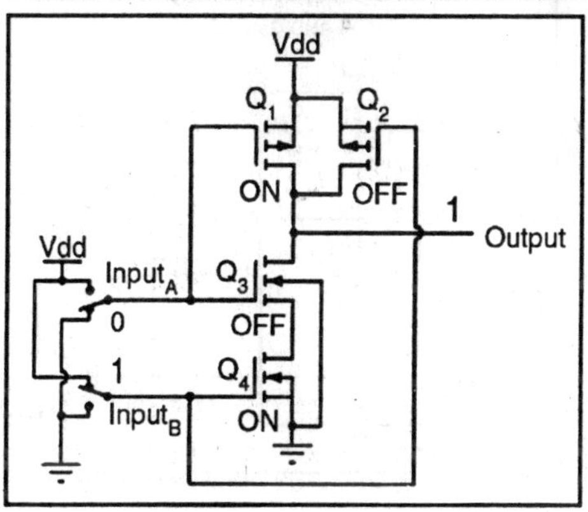

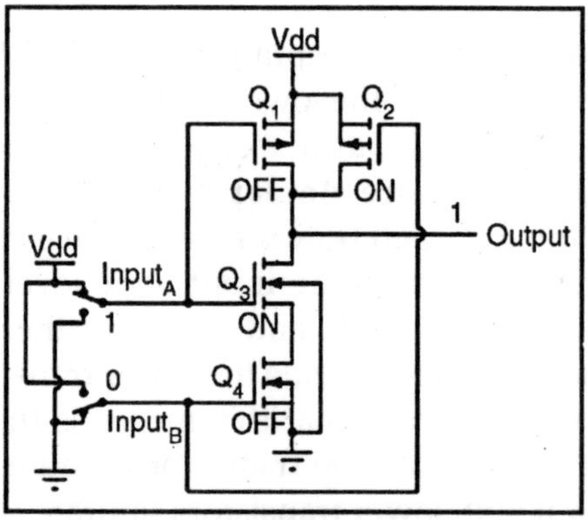


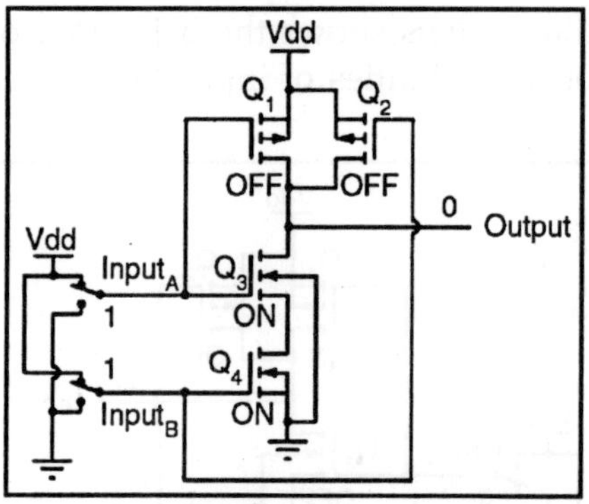


As with the TTL NAND gate, the CMOS NAND gate circuit may be used as the starting point for the creation of an AND gate. All that needs to be added is another stage of transistors to invert the output signal:

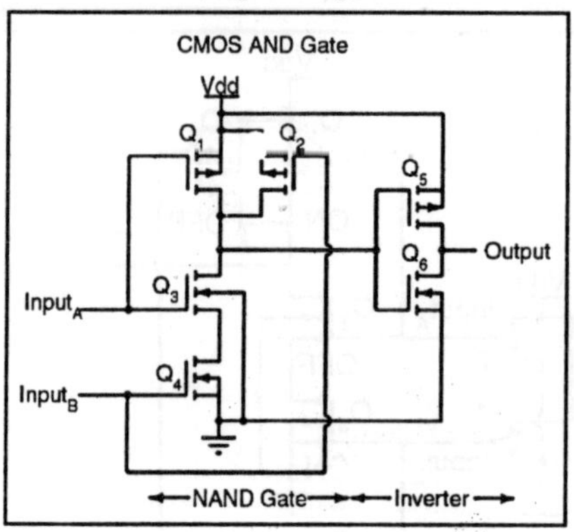

A CMOS NOR gate circuit uses four MOSFETs just like the NAND gate, except that its transistors are differently arranged. Instead of two paralleled sourcing (upper) transistors connected to $V_{dd}$ and two series-connected sinking (lower) transistors connected to ground, the NOR gate uses two series-connected sourcing transistors and two parallel-connected sinking transistors like this:

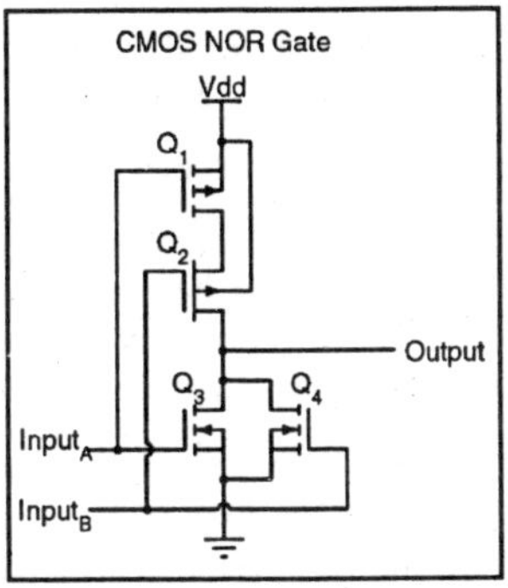


As with the NAND gate, transistors $Q_1$ and $Q_3$ work as a complementary pair, as do transistors $Q_2$ and $Q_4$. Each pair is controlled by a single input signal. If either input A or input B are "high" (1), at least one of the lower transistors ($Q_3$ or $Q_4$) will be saturated, thus making the output "low" (0). Only in the event of both inputs being "low" (0) will both lower transistors be in cutoff mode and both upper transistors be saturated, the conditions necessary for the output to go "high" (1). This behaviour, of course, defines the NOR logic function.

The OR function may be built up from the basic NOR gate with the addition of an inverter stage on the output:

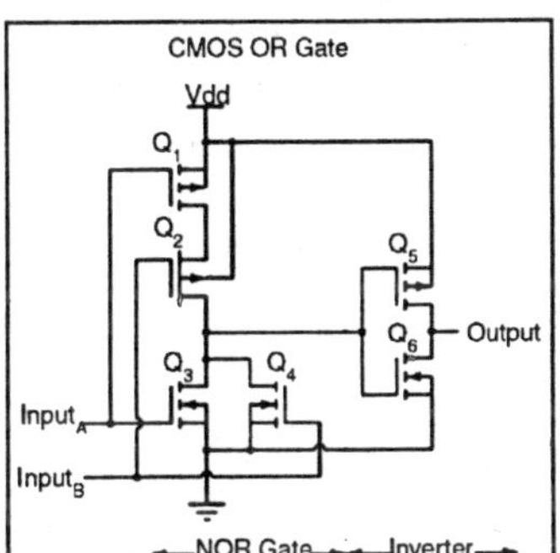

Since it appears that any gate possible to construct using TTL technology can be duplicated in CMOS, why do these two "families" of logic design still coexist? The answer is that both TTL and CMOS have their own unique advantages.

First and foremost on the list of comparisons between TTL and CMOS is the issue of power consumption. In this measure of performance, CMOS is the unchallenged victor.

Because the complementary P- and N-channel MOSFET pairs of a CMOS gate circuit are (ideally) never conducting at the same time, there is little or no current drawn by the circuit from the $V_{dd}$ power supply except for what current is necessary to source current to a load.

TTL, on the other hand, cannot function without some current drawn at all times, due to the biasing requirements of the bipolar transistors from which it is made.

There is a caveat to this advantage, though. While the power dissipation of a TTL gate remains rather constant regardless of its operating state(s), a CMOS gate dissipates more power as the frequency of its input signal(s) rises.

If a CMOS gate is operated in a static (unchanging) condition, it dissipates zero power (ideally). However, CMOS gate circuits draw transient current during every output state switch from "low" to "high" and visa-versa.

So, the more often a CMOS gate switches modes, the more often it will draw current from the $V_{dd}$ supply, hence greater power dissipation at greater frequencies.

A CMOS gate also draws much less current from a driving gate output than a TTL gate because MOSFETs are voltage-controlled, not current-controlled, devices.

This means that one gate can drive many more CMOS inputs than TTL inputs. The measure of how many gate inputs a single gate output can drive is called fanout.

Another advantage that CMOS gate designs enjoy over TTL is a much wider allowable range of power supply voltages.

Whereas TTL gates are restricted to power supply ($V_{cc}$) voltages between 4.75 and 5.25 volts, CMOS gates are typically able to operate on any voltage between 3 and 15 volts! The reason behind this disparity in power supply voltages is the respective bias requirements of MOSFET versus bipolar junction transistors.

MOSFETs are controlled exclusively by gate voltage (with respect to substrate), whereas BJTs are current-controlled devices.

TTL gate circuit resistances are precisely calculated for proper bias currents assuming a 5 volt regulated power supply.

Any significant variations in that power supply voltage will result in the transistor bias currents being incorrect, which then results in unreliable (unpredictable) operation.

The only effect that variations in power supply voltage have on a CMOS gate is the voltage definition of a "high" (1) state. For a CMOS gate operating at 15 volts of power supply voltage ($V_{dd}$), an input signal must be close to 15 volts in order to be considered "high" (1). The voltage threshold for a "low" (0) signal remains the same: near 0 volts.

One decided disadvantage of CMOS is slow speed, as compared to TTL. The input capacitances of a CMOS gate are much, much greater than that of a comparable TTL gate – owing to the use of MOSFETs rather than BJTs – and so a CMOS gate will be slower to respond to a signal transition (low-to-high or visa-versa) than a TTL gate, all other factors being equal.

The RC time constant formed by circuit resistances and the input capacitance of the gate tend to impede the fast rise- and fall-times of a digital logic level, thereby degrading high-frequency performance.

A strategy for minimizing this inherent disadvantage of CMOS gate circuitry is to "buffer" the output signal with additional transistor stages, to increase the overall voltage gain of the device.

This provides a faster-transitioning output voltage (high-to-low or low-to-high) for an input voltage slowly changing from one logic state to another. Consider this example, of an "unbuffered" NOR gate versus a "buffered," or B-series, NOR gate:

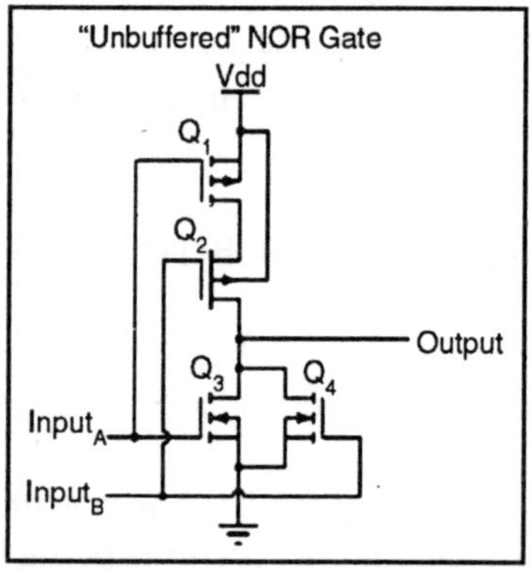


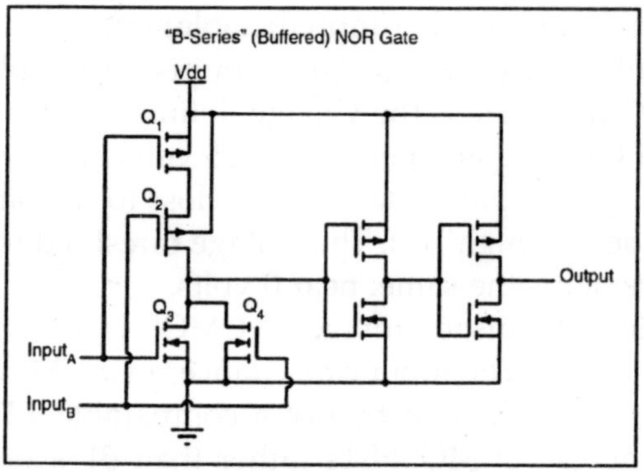


In essence, the B-series design enhancement adds two inverters to the output of a simple NOR circuit. This serves no purpose as far as digital logic is concerned, since two cascaded inverters simply cancel:

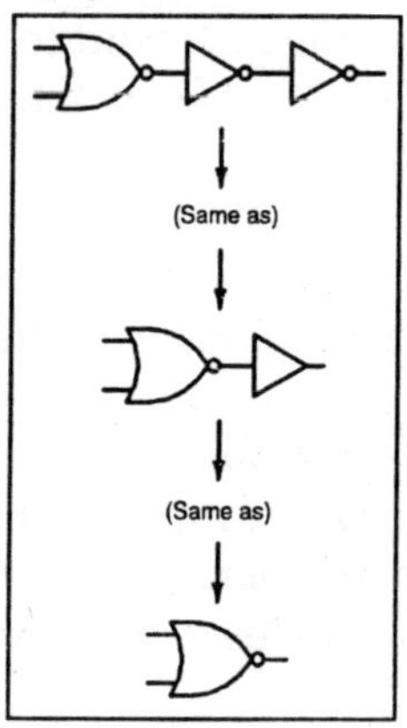

However, adding these inverter stages to the circuit does serve the purpose of increasing overall voltage gain, making the output more sensitive to changes in input state, working to overcome the inherent slowness caused by CMOS gate input capacitance.

*Review*:

- CMOS logic gates are made of IGFET (MOSFET) transistors rather than bipolar junction transistors.
- CMOS gate inputs are sensitive to static electricity. They may be damaged by high voltages, and they may assume any logic level if left floating.
- Pullup and pulldown resistors are used to prevent a CMOS gate input from floating if being driven by a signal source capable only of sourcing or sinking current.
- CMOS gates dissipate far less power than equivalent TTL gates, but their power dissipation increases with signal frequency, whereas the power dissipation of a TTL gate is approximately constant over a wide range of operating conditions.
- CMOS gate inputs draw far less current than TTL inputs, because MOSFETs are voltage-controlled, not current-controlled, devices.
- CMOS gates are able to operate on a much wider range of power supply voltages than TTL: typically 3 to 15 volts versus 4.75 to 5.25 volts for TTL.
- CMOS gates tend to have a much lower maximum operating frequency than TTL gates due to input capacitances caused by the MOSFET gates.
- B-series CMOS gates have "buffered" outputs to increase voltage gain from input to output, resulting in faster output response to input signal changes. This helps overcome the inherent slowness of CMOS gates due to MOSFET input capacitance and the RC time constant thereby engendered.

## SPECIAL-OUTPUT GATES

It is sometimes desirable to have a logic gate that provides both inverted and non-inverted outputs. For example, a single-input gate that is both a buffer and an inverter, with a separate output terminal for each function. Or, a two-input gate that provides both the AND and the NAND functions in a single circuit. Such gates do exist and they are referred to as complementary output gates. The general symbology for such a gate is the basic gate figure with a bar and two output lines protruding from it. An array of complementary gate symbols is shown in the following illustration:

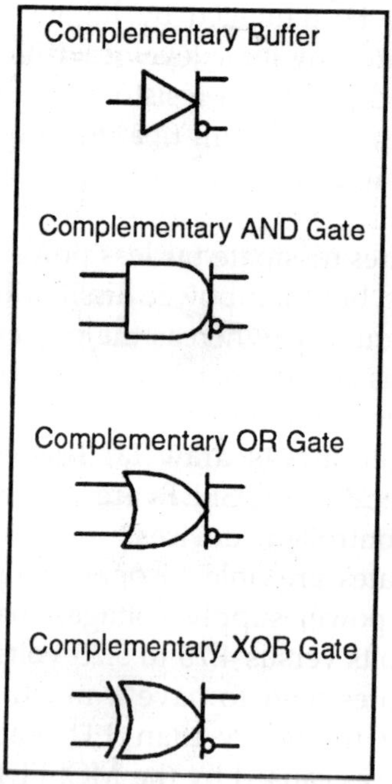


Complementary gates are especially useful in "crowded" circuits where there may not be enough physical room to mount the additional integrated circuit chips necessary to provide both inverted and noninverted outputs using standard gates and additional inverters. They are also useful in applications where a complementary output is necessary from a gate, but the addition of an inverter would introduce an unwanted time lag in the inverted output relative to the noninverted output. The internal circuitry of complemented gates is such that both inverted and noninverted outputs change state at almost exactly the same time:

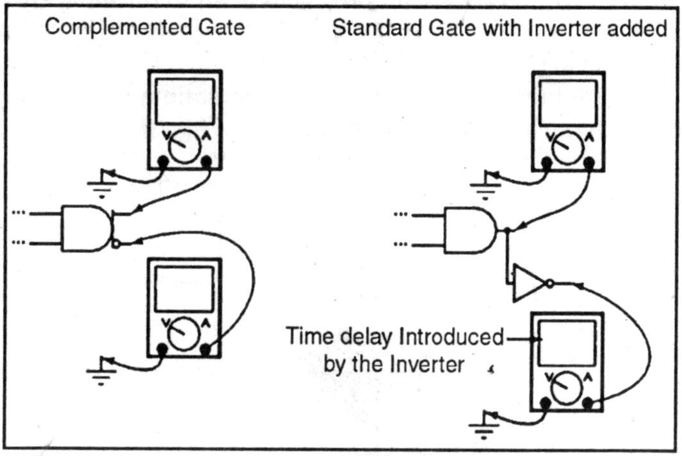


Another type of special gate output is called tristate, because it has the ability to provide three different output modes: current sinking ("low" logic level), current sourcing ("high"), and floating ("high-Z," or high-impedance). Tristate outputs are usually found as an optional feature on buffer gates. Such gates require an extra input terminal to control the "high-Z" mode, and this input is usually called the enable.

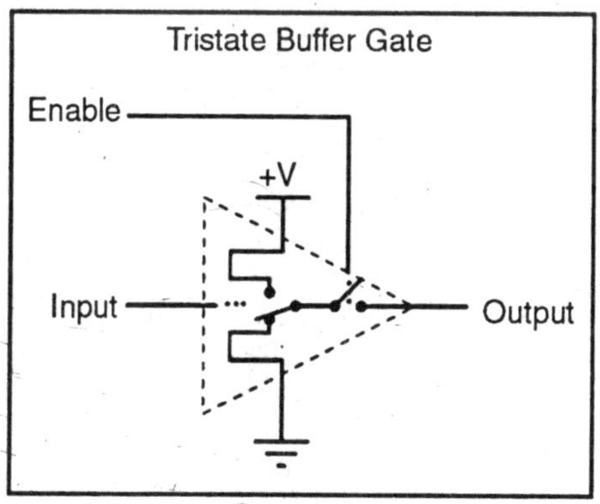


With the enable input held "high" (1), the buffer acts like an ordinary buffer with a totem pole output stage: it is capable of both sourcing and sinking current. However, the output terminal floats (goes into "high-Z" mode) if ever the enable input is grounded ("low"), regardless of the data signal's logic level. In other words, making the enable input terminal "low" (0) effectively disconnects the gate from whatever its output is wired to so that it can no longer have any effect.

Tristate buffers are marked in schematic diagrams by a triangle character within the gate symbol like this:

Tristate Buffer Symbol

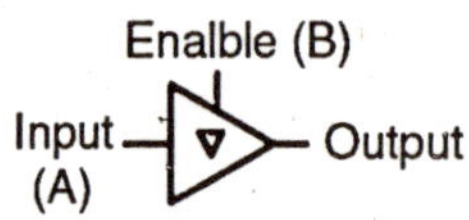


Truth Table

| A | B | Output |
|---|---|---|
| 0 | 0 | High-Z |
| 0 | 1 | 0 |
| 1 | 0 | High-Z |
| 1 | 1 | 1 |

Tristate buffers are also made with inverted enable inputs. Such a gate acts normal when the enable input is "low" (0) and goes into high-Z output mode when the enable input is "high" (1):

Tristate Buffer with Inverted Enable Input

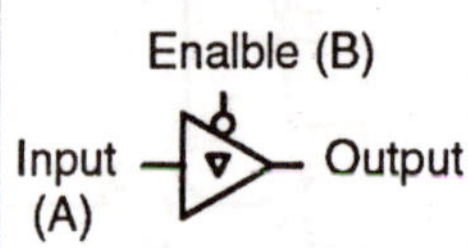


Truth Table

| A | B | Output |
|---|---|---|
| 0 | 0 | 0 |
| 0 | 1 | High-Z |
| 1 | 0 | 1 |
| 1 | 1 | High-Z |

One special type of gate known as the bilateral switch uses

Apollo space programme, CNC machine tool control computers, business ("mainframe") computers, and industrial control systems.

Despite the fact that core memory is long obsolete, the term "core" is still used sometimes with reference to a computer's RAM memory. All the while that delay lines, Williams Tube, and core memory technologies were being invented, the simple static RAM was being improved with smaller active component (vacuum tube or transistor) technology.

Static RAM was never totally eclipsed by its competitors: even the old ENIAC computer of the 1950's used vacuum tube ring-counter circuitry for data registers and computation. Eventually though, smaller and smaller scale IC chip manufacturing technology gave transistors the practical edge over other technologies, and core memory became a museum piece in the 1980's.

One last attempt at a magnetic memory better than core was the bubble memory. Bubble memory took advantage of a peculiar phenomenon in a mineral called garnet, which, when arranged in a thin film and exposed to a constant magnetic field perpendicular to the film, supported tiny regions of oppositely-magnetized "bubbles" that could be nudged along the film by prodding with other external magnetic fields.

"Tracks" could be laid on the garnet to focus the movement of the bubbles by depositing magnetic material on the surface of the film. A continuous track was formed on the garnet which gave the bubbles a long loop in which to travel, and motive force was applied to the bubbles with a pair of wire coils wrapped around the garnet and energized with a 2-phase voltage. Bubbles could be created or destroyed with a tiny coil of wire strategically placed in the bubbles' path. The presence of a bubble represented a binary "1" and the absence of a bubble represented a binary "0." Data could be read and written in this chain of moving magnetic bubbles as they passed by the tiny coil of wire, much the same as the read/write "head" in a cassette tape player, reading the magnetization of the tape as it moves.

Like core memory, bubble memory was nonvolatile: a permanent magnet supplied the necessary background field needed to support the bubbles when the power was turned off. Unlike core memory, however, bubble memory had phenomenal storage density: millions of bits could be stored on a chip of garnet only a couple of square inches in size. What killed bubble memory as a viable alternative to static and dynamic RAM was its slow, sequential data access.

Being nothing more than an incredibly long serial shift register (ring counter), access to any particular portion of data in the serial string could be quite slow compared to other memory technologies. An electrostatic equivalent of the bubble memory is the Charge-Coupled Device (CCD) memory, an adaptation of the CCD devices used in digital photography. Like bubble memory, the bits are serially shifted along channels on the substrate material by clock pulses.

Unlike bubble memory, the electrostatic charges decay and must be refreshed. CCD memory is therefore volatile, with high storage density and sequential access. Interesting, isn't it? The old Williams Tube memory was adapted from CRT viewing technology, and CCD memory from video recording technology.

## READ-ONLY MEMORY

Read-only memory (ROM) is similar in design to static or dynamic RAM circuits, except that the "latching" mechanism is made for one-time (or limited) operation. The simplest type of ROM is that which uses tiny "fuses" which can be selectively blown or left alone to represent the two binary states. Obviously, once one of the little fuses is blown, it cannot be made whole again, so the writing of such ROM circuits is one-time only. Because it can be written (programmed) once, these circuits are sometimes referred to as PROMs (Programmable Read-Only Memory).

However, not all writing methods are as permanent as blown fuses. If a transistor latch can be made which is resettable only with significant effort, a memory device that's something of a cross between a RAM and a ROM can be built.

Such a device is given a rather oxymoronic name: the EPROM (Erasable Programmable Read-Only Memory). EPROMs come in two basic varieties: Electrically-erasable (EEPROM) and Ultraviolet-erasable (UV/EPROM).

Both types of EPROMs use capacitive charge MOSFET devices to latch on or off. UV/EPROMs are "cleared" by long-term exposure to ultraviolet light. They are easy to identify: they have a transparent glass window which exposes the silicon chip material to light. Once programmed, you must cover that glass window with tape to prevent ambient light from degrading the data over time. EPROMs are often programmed using higher signal voltages than what is used during "read-only" mode.

## MEMORY WITH MOVING PARTS: "DRIVES"

The earliest forms of digital data storage involving moving parts was that of the punched paper card. Joseph Marie Jacquard invented a weaving loom in 1780 which automatically followed weaving instructions set by carefully placed holes in paper cards. This same technology was adapted to electronic computers in the 1950's, with the cards being read mechanically (metal-to-metal contact through the holes), pneumatically (air blown through the holes, the presence of a hole sensed by air nozzle backpressure), or optically (light shining through the holes).

An improvement over paper cards is the paper tape, still used in some industrial environments (notably the CNC machine tool industry), where data storage and speed demands are low and ruggedness is highly valued. Instead of wood-fibre paper, mylar material is often used, with optical reading of the tape being the most popular method. Magnetic tape (very similar to audio or video cassette tape) was the next logical improvement in storage media. It is still widely used today, as a means to store "backup" data for archiving and emergency restoration for other, faster methods of data storage.

Like paper tape, magnetic tape is sequential access, rather than random access. In early home computer systems, regular audio cassette tape was used to store data in modulated form, the binary 1's and 0's represented by different frequencies (similar to FSK data communication). Access speed was terribly slow (if you were reading ASCII text from the tape, you could almost keep up with the pace of the letters appearing on the computer's screen!), but it was cheap and fairly reliable.

Tape suffered the disadvantage of being sequential access. To address this weak point, magnetic storage "drives" with disk- or drum-shaped media were built. An electric motor provided constant-speed motion. A movable read/write coil (also known as a "head") was provided which could be positioned via servo-motors to various locations on the height of the drum or the radius of the disk, giving access that is almost random (you might still have to wait for the drum or disk to rotate to the proper position once the read/write coil has reached the right location).

The disk shape lent itself best to portable media, and thus the floppy disk was born. Floppy disks (so-called because the magnetic media is thin and flexible) were originally made in 8-inch diameter formats. Later, the 5-1/4 inch variety was introduced, which was made practical by advances in media particle density.

All things being equal, a larger disk has more space upon which to write data. However, storage density can be improved by making the little grains of iron-oxide material on the disk substrate smaller. Today, the 3-1/2 inch floppy disk is the preeminent format, with a capacity of 1.44 Mbytes (2.88 Mbytes on SCSI drives). Other portable drive formats are becoming popular, with IoMega's 100 Mbyte "ZIP" and 1 Gbyte "JAZ" disks appearing as original equipment on some personal computers.

Still, floppy drives have the disadvantage of being exposed to harsh environments, being constantly removed from the drive mechanism which reads, writes, and spins the media. The first disks were enclosed units, sealed from all dust and other particulate matter, and were definitely not portable.

Keeping the media in an enclosed environment allowed engineers to avoid dust altogether, as well as spurious magnetic fields. This, in turn, allowed for much closer spacing between the head and the magnetic material, resulting in a much tighter-focused magnetic field to write data to the magnetic material.

The following photograph shows a hard disk drive "platter" of approximately 30 Mbytes storage capacity. A ball-point pen has been set near the bottom of the platter for size reference:

Modern disk drives use multiple platters made of hard material (hence the name, "hard drive") with multiple read/write heads for every platter. The gap between head and platter is much smaller than the diameter of a human hair. If the hermetically-sealed environment inside a hard disk drive is contaminated with outside air, the hard drive will be rendered useless. Dust will lodge between the heads and the platters, causing damage to the surface of the media.

Here is a hard drive with four platters, although the angle of the shot only allows viewing of the top platter. This unit is complete with drive motor, read/write heads, and associated electronics. It has a storage capacity of 340 Mbytes, and is about the same length as the ball-point pen shown in the previous photograph:

While it is inevitable that non-moving-part technology will replace mechanical drives in the future, current state-of-the-art electromechanical drives continue to rival "solid-state" nonvolatile memory devices in storage density, and at a lower cost.

In 1998, a 250 Mbyte hard drive was announced that was approximately the size of a quarter (smaller than the metal platter hub in the centre of the last hard disk photograph)! In any case, storage density and reliability will undoubtedly continue to improve.

An incentive for digital data storage technology advancement was the advent of digitally encoded music. A joint venture between Sony and Phillips resulted in the release of the "compact audio disk" (CD) to the public in the late 1980's. This technology is a read-only type, the media being a thin film of aluminum foil embedded in a transparent plastic disk.

Binary bits are "burned" into the aluminum as pits by a high-power laser. Data is read by a low-power laser (the beam of which can be focused more precisely than normal light) reflecting off the aluminum to a photocell receiver. The advantages of CDs over magnetic tape are legion. Being digital, the information is highly resistant to corruption. Being non-contact in operation, there is no wear incurred through playing.

Being optical, they are immune to magnetic fields (which can easily corrupt data on magnetic tape or disks). It is possible to purchase CD "burner" drives which contain the high-power laser necessary to write to a blank disk. Following on the heels of the music industry, the video entertainment industry has leveraged the technology of optical storage with the introduction of the Digital Video Disk, or DVD.

Using a similar-sized plastic disk as the music CD, a DVD employs closer spacing of pits to achieve much greater storage density. This increased density allows feature-length movies to be encoded on DVD media, complete with trivia information about the movie, director's notes, and so on. Much effort is being directed toward the development of practical read/write optical disks (CD-W).

# PLA

## INTRODUCTION

One way to design a combinational logic circuit it to get gates and connect them with wires. One disadvantage with this way of designing circuits is its lack of portability. You can now get chips called PLA (programmable logic arrays) and "programme" them to implement Boolean functions. I'll explain what it means to programme a PLA. Fortunately, a PLA is quite simple to learn, and produces nice neat circuits too.

## STARTING OUT

*The first part of a PLA looks like:*

Each variable is hooked to a wire, and to a wire with a NOT gate. So the top wire is $x_2$ and the one just below is its negation, \$x_2$. Then there's x1 and just below it, its negation, \$x_1$. The next part is to draw a vertical wire with an AND gate. I've drawn 3 of them.

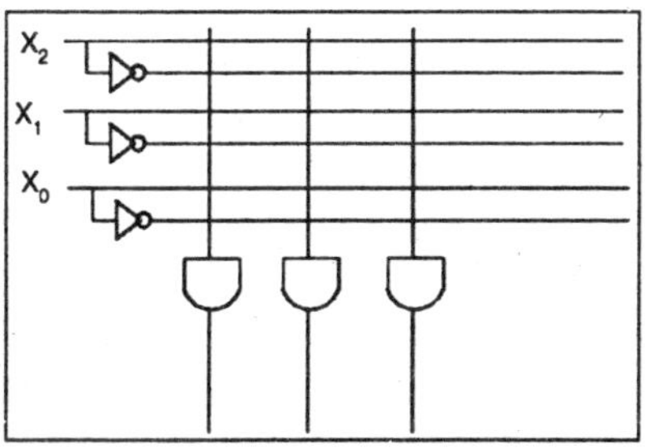


Let's try to implement a truth table with a PLA.

| $x_2$ | $x_1$ | $x_0$ | $z_1$ | $z_0$ |
|---|---|---|---|---|
| 0 | 0 | 0 | 0 | 0 |
| 0 | 0 | 1 | 1 | 0 |

| | | | | |
|---|---|---|---|---|
| 0 | 1 | 0 | 0 | 0 |
| 0 | 1 | 1 | 1 | 0 |
| 1 | 0 | 0 | 1 | 1 |
| 1 | 0 | 1 | 0 | 0 |
| 1 | 1 | 0 | 0 | 0 |
| 1 | 1 | 1 | 0 | 1 |

Each of the vertical lines with an AND gate corresponds to a minterm. For example, the first AND gate (on the left) is the minterm: $\backslash x_2 \backslash x_1 x_0$.

- The second AND gate (from the left) is the minterm: $\backslash x_2 x_1 x_0$.
- The third AND gate (from the left) is the minterm: $x_2 \backslash x_1 \backslash x_0$.
- I've added a fourth AND gate which is the minterm: $x_2 x_1 x_0$.
- The first three minterms are used to implement $z_1$. The third and fourth minterm are used to implement z0.

This is how the PLA looks after we have all four minterms.

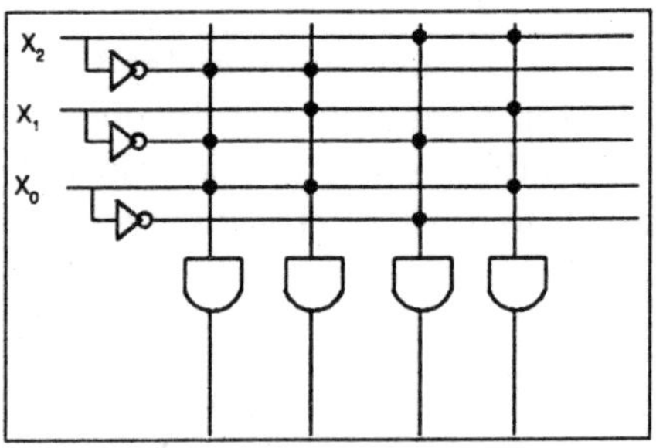


Now you might complain. How is it possible to have a one input AND gate? How can three inputs be hooked to the same wire to an AND gate? Isn't that invalid for combinational logic circuits?

That's true, it is invalid. However, the diagram is merely a simplification. I've drawn the each of AND gate with three input wires, which is what it is in reality (there is as many input wires as variables). For each connection (shown with a black dot), there's really a separate wire. We draw one wire just to make it look neat.

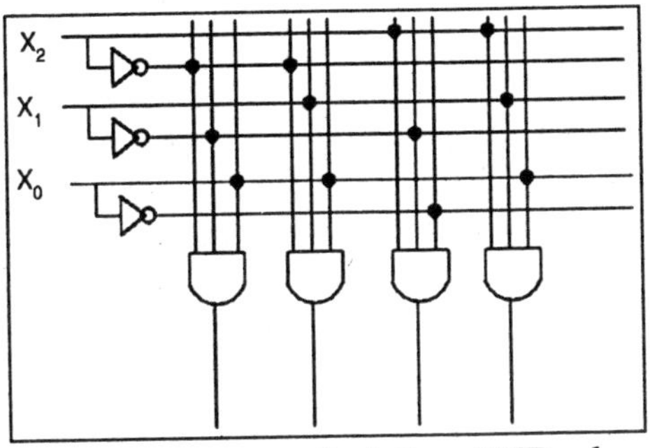


The vertical wires are called the AND plane. We often leave out the AND gates to make it even easier to draw.

We then add OR gates using horizontal wires, to connect the minterms together.

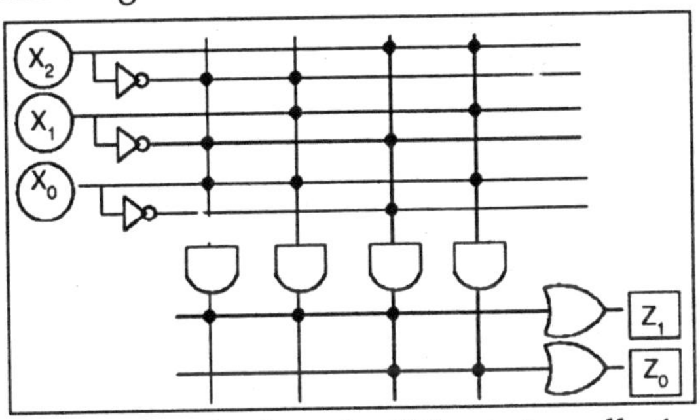


Again, a single wire into the OR gate is really 4 wires. We use the same simplification to make it easier to read. The horizontal wires make up the OR plane. This is how the PLA looks when we leave out the AND gates and the OR gates. It's not that the AND gates and OR gates aren't there---they are, but they've been left out to make the PLA even easier to draw.

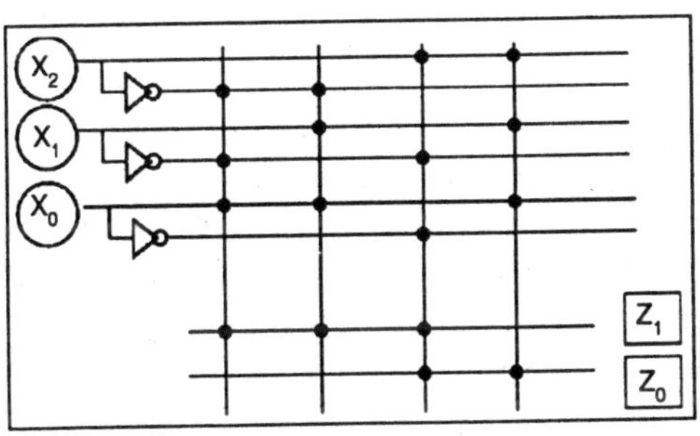

## MINTERMS

Given n variables, it would seem necessary to have $2^n$ vertical wires (for the AND gates), one for each possible minterm. However, $2^n$ grows VERY quickly. So, sometimes there aren't $2^n$ vertical wires. You can generally get around the problem by not connecting the wire to each of the three variables. For example, you could just have a product term $x_2 \backslash x_0$ or even simply $\backslash x_1$.

## PAL

The term Programmable Array Logic (PAL) is used to describe a family of programmable logic device semiconductors used to implement logic functions in digital circuits introduced by Monolithic Memories, Inc. (MMI) in March 1978.

The PAL device is a special case of PLA which has a programmable AND array and a fixed OR array. The basic structure of Rom is same as PLA. It is cheap compared to PLA as only the AND array is programmable. It is also easy to programme a PAL compared to PLA as only AND must be programmed.

PAL devices consisted of a small PROM (programmable read-only memory) core and additional output logic used to implement particular desired logic functions with few components. Using specialized machines, PAL devices were "field-programmable". Each PAL device was "one-time programmable" (OTP), meaning that it could not be updated and reused after its initial programming. (MMI also offered a similar family called HAL, or "hard array logic", which were like PAL devices except that they were mask-programmed at the factory.)

The figure below shows a segment of an unprogrammed PAL. The input buffer with non inverted and inverted outputs is used, since each PAL must drive many AND Gates inputs. When the PAL is programmed, the fusible links (F1, F2, F3...F8) are selectively blown to leave the desired connections to the AND Gate inputs. Connections to the AND Gate inputs in a PAL are represented by Xs, as shown here:

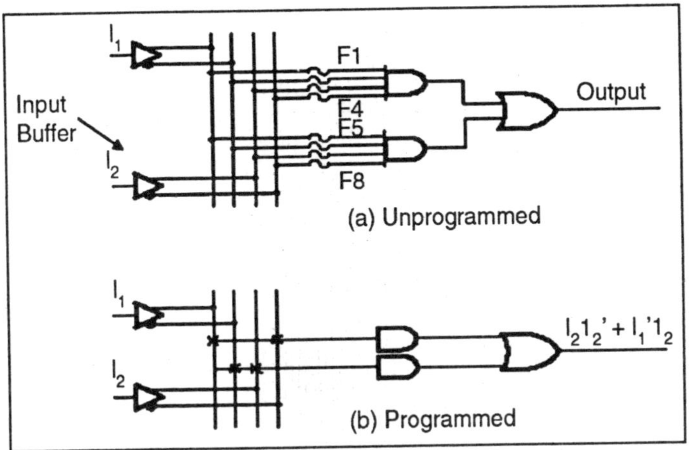


Typical combinational PAL have 10 to 20 inputs and from 2 to 10 outputs with 2 to 8 AND gates driving each OR gate. PALs are also available which contain D flip-flops with inputs driven from the programming array logic. Such PAL provides a convenient way of realizing sequential networks.

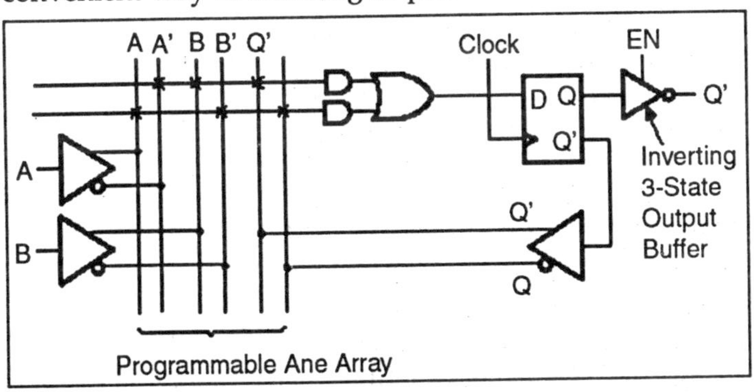


**Fig.** Below Shows a Segment of a Sequential PAL.

The D flip-flop is driven from the OR gate, which is fed by two AND gates. The flip-flop output is fed back to the programmable AND array through a buffer. Thus the AND gate inputs can be connected to A, A', B, B', Q, or Q'. The Xs on the diagram show the realization of the next-state equation.

$$Q+ = D = A'BQ' + AB'Q$$

The flip-flop output is connected to an inverting tristate buffer, which is enabled when,

EN = 1

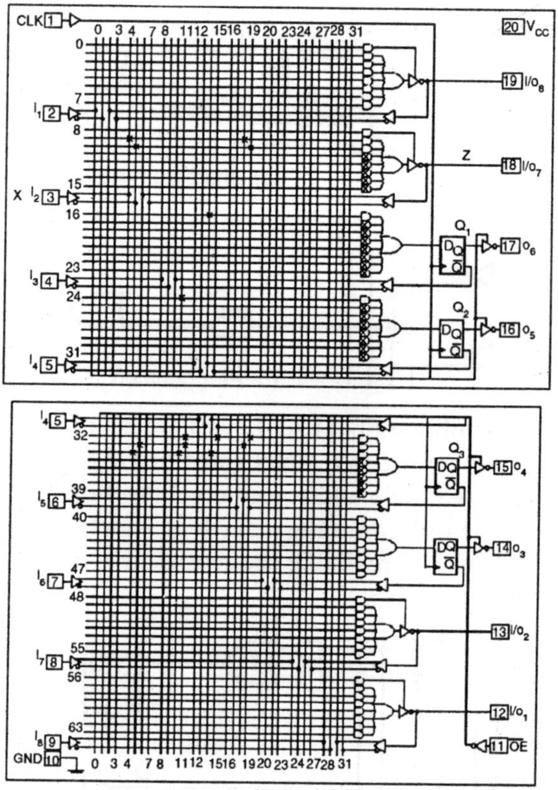

**Fig.** Shows a Logic Diagram for
a Typical Sequential PAL, the 16R4

This PAL has an AND gate array with 16 input variables, and it has 4 D flip-flops. Each flip-flop output goes through a tristate-inverting buffer (output pins 14-17). One input (pin 11) is used to enable these buffers. The rising edge of a common clock (pin 1) causes the flip-flops to change the state. Each D flip-flop input is driven from an OR gate, and each OR gate is fed from 8 AND gates.

The AND gate inputs can come from the external PAL inputs (pins2-9) or from the flip-flop outputs, which are fed back internally. In addition there are four input/output (i/o) terminals (pins 12,13,18 and 19), which can be used as either network outputs or as inputs to the AND gates. Thus each AND gate can have a maximum of 16 inputs (8 external inputs, 4 inputs fed back from the flip-flop outputs, and 4 inputs from the i/o terminals). When used as an output, each I/O terminal is driven from an inverting tristate buffer. Each of these buffers is fed from an OR gate and each OR gate is fed from 7 AND gates. An eighth AND gate is used to enable the buffer.

## PAL ARCHITECTURE

The PAL architecture consists of two main components: a logic plane and output logic macrocells.

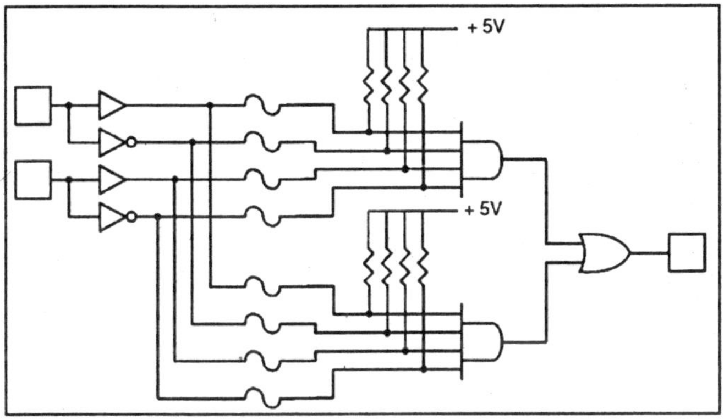


Fig. Simplified Programmable Logic Device

### Programmable Logic Plane

The programmable logic plane is a programmable (PROM) array that allows the signals present on the devices pins (or the logical complements of those signals) to be routed to an output logic macrocell.

PAL devices have arrays of transistor cells arranged in a "fixed-OR, programmable-AND" plane used to implement "sum-of-products" binary logic equations for each of the outputs in terms of the inputs and either synchronous or asynchronous feedback from the outputs.

# Index